MUTAGENICITY TESTING IN ENVIRONMENTAL POLLUTION CONTROL

ELLIS HORWOOD SERIES IN ANALYTICAL CHEMISTRY

Series Editors: Dr. R. A. CHALMERS and Dr. MARY MASSON
University of Aberdeen

APPLICATION OF ION SELECTIVE MEMBRANE ELECTRODES IN ORGANIC ANALYSIS
G. E. BAIULESCU and V. V. COSOFRET
EDUCATION AND TEACHING IN ANALYTICAL CHEMISTRY
G. E. BAIULESCU, C. PATROESCU and R. A. CHALMERS
HANDBOOK OF PRACTICAL ORGANIC MICROANALYSIS
S. BANCE
ORGANIC TRACE ANALYSIS
K. BEYERMANN
FOUNDATIONS OF CHEMICAL ANALYSIS
O. BUDEVSKY
INORGANIC REACTION CHEMISTRY Volume 1: Systematic Chemical Separation
D. T. BURNS, A. TOWNSHEND and A. G. CATCHPOLE
INORGANIC REACTION CHEMISTRY Reactions of the Elements and their Compounds Volume 2, Part A: Alkali Metals to Nitrogen and Volume 2, Part B: Osmium to Zirconium
D. T. BURNS, A. TOWNSHEND and A. H. CARTER
IMPROVED HOLLOW CATHODE LAMPS FOR ATOMIC SPECTROSCOPY
S. CAROLI
AUTOMATIC CHEMICAL ANALYSIS
J. K. FOREMAN and P. B. STOCKWELL
FUNDAMENTALS OF ELECTROCHEMICAL ANALYSIS
Z. GALUS
LABORATORY HANDBOOK OF PAPER AND THIN LAYER CHROMATOGRAPHY
J. GASPARIČ and J. CHURÁČEK
HANDBOOK OF ANALYTICAL CONTROL OF IRON AND STEEL PRODUCTION
T. S. HARRISON
ORGANIC ANALYSIS USING ATOMIC ABSORPTION SPECTROMETRY
SAAD S. M. HASSAN
HANDBOOK OF ORGANIC REAGENTS IN INORGANIC CHEMISTRY
Z. HOLZBECHER, L. DIVIŠ, M. KRÁL, L. ŠŮCHA and F. VLÁČIL
GENERAL HANDBOOK OF ON-LINE PROCESS ANALYSERS
DAVID HUSKINS
QUALITY MEASURING INSTRUMENTS IN ON-LINE PROCESS ANALYSIS
DAVID HUSKINS
ANALYTICAL APPLICATIONS OF COMPLEX EQUILIBRIA
J. INCZÉDY
PARTICLE SIZE ANALYSIS
Z. K. JELÍNEK
OPERATIONAL AMPLIFIERS IN CHEMICAL INSTRUMENTATION
R. KALVODA
METHODS OF PROTEIN ANALYSIS
Edited by I. KERESE
HANDBOOK OF CHEMICAL EQUILIBRIA IN ANALYTICAL CHEMISTRY
S. KOTRLÝ and L. ŠŮCHA
ATLAS OF METAL-LIGAND EQUILIBRIA IN AQUEOUS SOLUTION
J. KRAGTEN
GRADIENT LIQUID CHROMATOGRAPHY
C. LITEANU and S. GOCAN
STATISTICAL THEORY AND METHODOLOGY OF TRACE ANALYSIS
C. LITEANU and I. RÎCĂ
SPECTROPHOTOMETRIC DETERMINATION OF ELEMENTS
Z. MARCZENKO
LABORATORY HANDBOOK OF CHROMATOGRAPHIC AND ALLIED METHODS
O. MIKEŠ
STATISTICS FOR ANALYTICAL CHEMISTRY
J. N. MILLER and J. C. MILLER
SEPARATION AND PRECONCENTRATION METHODS IN INORGANIC TRACE ANALYSIS
J. MINCZEWSKI, J. CHWASTOWSKA and R. DYBCZYNSKI
HANDBOOK OF ANALYSIS OF ORGANIC SOLVENTS
V. ŠEDIVEC and J. FLEK
ELECTROCHEMICAL STRIPPING ANALYSIS
F. VYDRA, K. ŠTULÍK and B. JULÁKOVÁ
ANALYSIS WITH ION-SELECTIVE ELECTRODES
J. VESELÝ, D. WEISS and K. ŠTULÍK
MUTAGENICITY TESTING IN ENVIRONMENTAL POLLUTION CONTROL
F. K. ZIMMERMANN and R. E. TAYLOR-MAYER

MUTAGENICITY TESTING IN ENVIRONMENTAL POLLUTION CONTROL

F. K. ZIMMERMANN
Institute für Mikrobiologie, Technische Hochschule
Darmstadt, West Germany

and

R. E. TAYLOR-MAYER
Slippery Rock University, Pennsylvania, USA

ELLIS HORWOOD LIMITED
Publishers · Chichester

Halsted Press: a division of
JOHN WILEY & SONS
New York · Chichester · Brisbane · Toronto

First published in 1985 by
ELLIS HORWOOD LIMITED
Market Cross House, Cooper Street, Chichester, West Sussex, PO19 1EB, England

The publisher's colophon is reproduced from James Gillison's drawing of the ancient Market Cross, Chichester.

Distributors:

Australia, New Zealand, South-east Asia:
Jacaranda-Wiley Ltd., Jacaranda Press,
JOHN WILEY & SONS INC.,
G.P.O. Box 859, Brisbane, Queensland 4001, Australia

Canada:
JOHN WILEY & SONS CANADA LIMITED
22 Worcester Road, Rexdale, Ontario, Canada.

Europe, Africa:
JOHN WILEY & SONS LIMITED
Baffins Lane, Chichester, West Sussex, England.

North and South America and the rest of the world:
Halsted Press: a division of
JOHN WILEY & SONS
605 Third Avenue, New York, N.Y. 10158 U.S.A.

British Library Cataloguing in Publication Data
Zimmermann, F.K.
Mutagenicity testing in environmental pollution control. –
(Ellis Horwood series in analytical chemistry)
1. Mutagenicity testing 2. Pollutants – Testing
I. Title II. Taylor-Mayer, R.E.
628.5 QH465

ISBN 0–85312–681–X (Ellis Horwood Limited)
ISBN 0–470–20216–5 (Halsted Press)

Printed in Great Britian by Butler & Tanner, Frome, Somerset

Table of Contents

Foreword

The screening of environmental chemicals for potential mutagenic and carcinogenic activity in the human population continues to be a high-priority activity all over the world. The screening type experiments that were started in the late 1960s clearly showed that there were chemicals with mutagenic activity in every class of chemicals analysed. This screening has continued and concomitantly there has been the development of new approaches and new assay systems designed to detect mutagenic/carcinogenic activity in foods, beverages and other formulated products. Much of the new research is concerned with the evaluation of complex mixtures, both as they occur as air pollutants or pollutants of fresh and sea water. Each of these latter situations offer a unique challenge; the pollutants are complex mixtures and highly variable in content from one geographical location to another. A particularly imaginative approach is to use assays on indigenous animal and plant species as they can provide an assessment of pollutants as they occur in the environment. In other cases it is possible to introduce assay systems such as maize or Tradescantia into an ecosystem to provide a mechanism for evaluation of air pollutants, for example, where there are no known indigenous species that have been studied sufficiently to be exploited in this same manner.

Monitoring the environment and developing new approaches and techniques to collect data to make it possible to evaluate the human health consequences of exposure to environmental chemicals continues to offer a challenge to the scientific community. Much progress has been made in the area of genetic toxicology and environmental mutagenesis and carcinogenesis as discussed in the present book. We are, however, pretty much limited to evaluation of the effects on experimental organisms, and we need to expand this capability to man himself to determine the long-term human health effects. The research discussed in this book brings us another step closer to this type of assessment and reminds the reader that a problem with great significance for the human population has been identified. May this book serve as a stimulus to even greater creative thinking and test-method development that will enable us to take the final step – that of risk assessment. It is only when we can accurately assess the impact of exposure to environmental chemical pollutants on human health – both with regard to effects on the exposed population as well as its progeny – that this cycle of research can be considered complete.

Frederick J. de Serres

1

Introduction

F. K. ZIMMERMANN,
Technische Hochschule, Darmstadt, FRG

Our environment contains a vast variety of agents which are detrimental to human health. Agents which are acutely toxic are readily detected because the immediate effects allow rapid identification of the source of toxicity. Mutations or tumours are not the immediate result of an exposure to mutagenic or carcinogenic agents. Rather, they are only detected many years after exposure. The latent period for tumour induction in humans may be as long as 20 years. Recessive mutations can only be detected after several generations. A source of mutagenic or carcinogenic activity may only be detected after many persons have been exposed for many years. Therefore, preventive action is called for. Our natural and industrial environment has to be monitored for mutagenic and carcinogenic pollution. Short-term tests are now available and used to achieve this goal. Of course, we are only at the beginning, but a number of promising approaches have been used, and it is the goal of this small book to provide a few examples of what is being tried.

Testing for mutagenic activity is used not only for the detection of mutagens, as it has turned out that mutagenicity tests can be used as short-term carcinogenicity tests. Mutation assays are shorter and much less costly than classical carcinogenicity tests, and thus it has become feasible to develop systems for the routine control of mutagenic/carcinogenic environmental pollution.

We already know numerous mutagenic and carcinogenic chemicals, and it may appear that simple tests by analytical chemistry should suffice to screen environmental samples for mutagenic/carcinogenic pollution. However, the numbers of such chemicals are very high and new ones are constantly reported and even more may have evaded identification. An appropriate biological test battery will monitor mutagenic/carcinogenic pollution by known as well as unknown agents. Therefore, biological mutation assays are indispensable tools in environmental pollution control.

Human health is our main concern, and the first contribution by Vig is intended to provide a brief overview of the genetic apparatus of humans. The second contribution by Zimmermann and Taylor-Mayer provides some basic information on mutation and changes induced in the genetic apparatus of cells, as well as some background on actual mutagenicity testing.

One major problem with environmental samples is that they are complex chemical mixtures. The different components can interfere with each other so that the mutagenic components are masked. Therefore, fractionation of such samples is important and Møller and Bjørseth report on an elegant approach to combining fractionation with a coupled mutation assay.

The public at large is mainly concerned with pollution caused by industrial chemicals, and there is a tendency to consider our 'natural' environment as safe. Nagao and his colleagues report on their work on the identification of mutagenic activities in our 'normal' diets, food and drink. These results may lead in the long run to changes in our eating and drinking habits as well as in our ways of preparing food and beverages.

Water has become a precious commodity world wide, and its quality has became a major concern because of rampant environmental pollution. Water quality has not only to be monitored for high load of salts but also for the presence of toxic and even mutagenic/carcinogenic material. Van Kreijl and Sloof report on their efforts to establish such a monitoring system for Dutch river waters. They not only test water extracts for mutagenic activity; they are also trying to use fish as monitoring organisms in their natural environment. Parry is concerned with mutagenic pollution of the marine environment. Normally, many dangerous chemicals are present in very low concentrations in water bodies, but they can be concentrated again in the food chain. Mussels are such 'concentrators' and Parry has successfully used extracts from mussels collected at different sites to follow mutagenic pollution along the coast of southern Wales.

Nestmann reports on a project to identify mutagenic pollution in paper-mill effluents. He also documents the necessity for a test battery rather than a single mutation assay. A reliable detection of mutagenic pollution is not possible on a single-test basis.

Combustion is a notorious source of highly dangerous pollution, and air quality has also to be monitored in respect to the presence of mutagenic pollution. Alink and Van Houdt report on their efforts to develop a system for the detection of mutagenic air pollution.

We have to make sure that our produce is free from mutagenic pollution. This goal can only be achieved if the basis of our agriculture is controlled in respect to mutagenic contamination. Plewa has developed plant systems to detect mutagenic activity directly in the field. His results clearly indicate that monitoring at this level is very urgent.

There has been a tendency to employ mutation assays as they have become available and advertised rather than setting up test batteries on the basis of genetic merits. Zimmermann *et al.* report on results obtained with a test specifically designed to detect induction of chromosomal mal-segregation and have been able to show that there are indeed chemicals which are mutagenic only in respect to the induction of aneuploidy but negative in all other tests used.

An immense volume of mutagenicity test data has been accumulated during the last 40 years, and the question is what to do with it. What is needed now is a risk evaluation. This is very difficult. However, as discussed by Unrau, there is a precedent. risk in respect to the induction of mutation and cancer has been evaluated for ionizing radiation, and national as well as international authorities have set exposure limits. It may not be possible to extrapolate directly from ionizing radiation to chemicals, but the fact that a risk evaluation has been established may provide guidelines along which to develop recommendations for certain chemical pollutants.

2

The genetic apparatus of man

BALDEV K. VIG, Department of Biology, University of Nevada, Reno, Nevada 89557, USA

2.1 INTRODUCTION

In 1956 Tjio and Levan showed that the correct somatic chromosome number of man is 46. This discovery, made by the use of colchicine and a hypotonic solution, revolutionized human genetics. Not only did it help establish variations in chromosome number and structure in many human disorders, but it also stimulated research on the organization, replication and physiochemical nature of the genetic material of man. If Garrod's thesis on "Inborn Errors of Metabolism" was an embryo for the field of human genetics, the methodology of human cytogenetics resulted in the birth of a formal discipline.

The haploid set of human chromosomes found in the sperm or egg nucleus contains a total of 1.5 pg (1100mm) of DNA packed into 23 discrete units. After fertilization the diploid cell undergoes physical changes. The chromosomal material appears as fine filamentous, intertwined structures which cannot be resolved into individual units. This mass, called chromatin, is surrounded by a double-layered nuclear membrane. Outside the nucleus the cytoplasm is loaded with mitochondria, golgi apparatus, polyribosomal complexes, and messenger, transfer and ribosomal RNA of maternal origin, along with other molecules necessary for the survival, function and multiplication of the cell.

2.2 MITOTIC CELL DIVISION

The zygote, as well as other somatic cells, duplicates by a mechanism which produces two genetically identical cells. This process, termed *mitosis,* is traditionally divided into several stages. Between cycles of division the cell is said to be in the *interphase.* In the G_1 period of interphase, lasting 3–4 hours, a variety of proteins is synthesized. G_1 is followed by a period of DNA synthesis, the S phase, which is of 6–8 hour duration in most eukaryotes. Proteins of various types are also synthesized in the S phase. Another burst of protein synthesis follows during a 3–4 hour period called G_2.

As the actual division process begins with the *prophase,* the chromatin appears as long thread-like structures which continue to condense and shorten into the next phase, the *metaphase.* It is during the metaphase that cytogeneticists study the numerical and structural details of karyotype.

At this stage each of the 46 human chromosomes is composed of two strands called *chromatids,* the parallel halves of a chromosome. The chromatids are held together at only one point, called the centromere, kinetochore, or primary constriction. A spindle apparatus, consisting of microtubules, forms during the prophase and in mid to late metaphase all chromosomes line up around the spindle with their centromeres characteristically along the equatorial line of the cell. The nuclear membrane breaks into small pieces which disintegrate.

The centromeres are attached to spindle fibres from both poles of the spindle. As each centromere divides into two units, one unit or daughter kinetochore is pulled toward one pole of the spindle and the other kinetochore to the other pole. This poleward movement, which occurs during *anaphase,* divides 46 double-stranded chromosomes into two sets of 46 single-stranded structures, one set moving to one pole and the other set of sister structures moving to the other pole. At the two poles each of the sets, now in *telophase,* go through a reverse sequence of events. The spindle fibres disintegrate, the chromosomes decondense and become surrounded by newly formed nuclear envelopes, and the cell cytoplasm divides into two roughly equal parts.

Two equivalent cells are formed from one. Each of these cells, because it carries the entire genetic apparatus present in the original zygote, is called *totipotential.* The division of each of these two daughter cells follows the same pattern and about 50 cell divisions (each taking roughly 18–20 hours) result in an individual consisting of some 10^{13}–10^{15} cells. This number of cell divisions is sufficient to allow cells to specialize or differentiate, hence no further mitosis or equational division occurs in some organs.

2.3 THE HUMAN GENOME

At mitotic metaphase the chromosomes of man appear as shown in Fig. 2.1. Twenty-three are of paternal origin; the other 23 are the maternal contribution. Therefore, one human genome has two homologues of each type of chromosome. It is difficult or even impossible to distinguish between some chromosome pairs (e.g. numbers 13, 14 and 15). However, when such metaphase chromosomes are exposed to dilute solutions of trypsin, or certain other treatments, each pair shows characteristic banding patterns. On this basis all chromosome pairs can be identified. The karyotype in Fig. 2.2 is an example of chromosomes arranged in their traditional positions. This cell, derived from a male, has 46 chromosomes, including the two *sex chromosomes,* X and Y. The female has

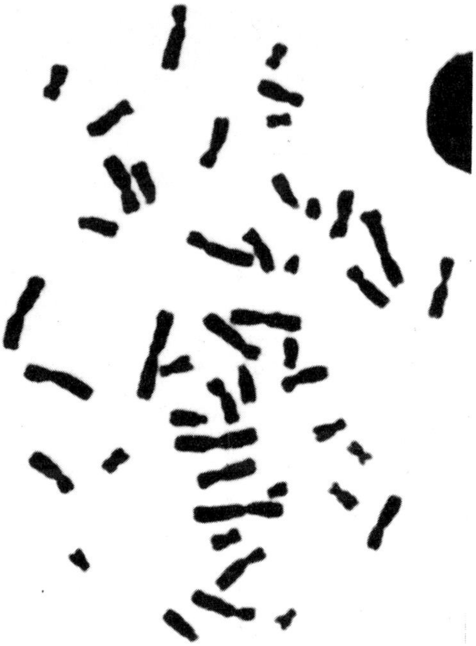

Fig. 2.1 – A male human cell at metaphase, showing 46 chromosomes. This preparation was made from lymphocyte cultures. The cells were arrested at metaphase with colcemid and treated with hypotonic solution. Note that every chromosome has two chromatids. Because of the action of colcemid the spindle microtubules are not formed.

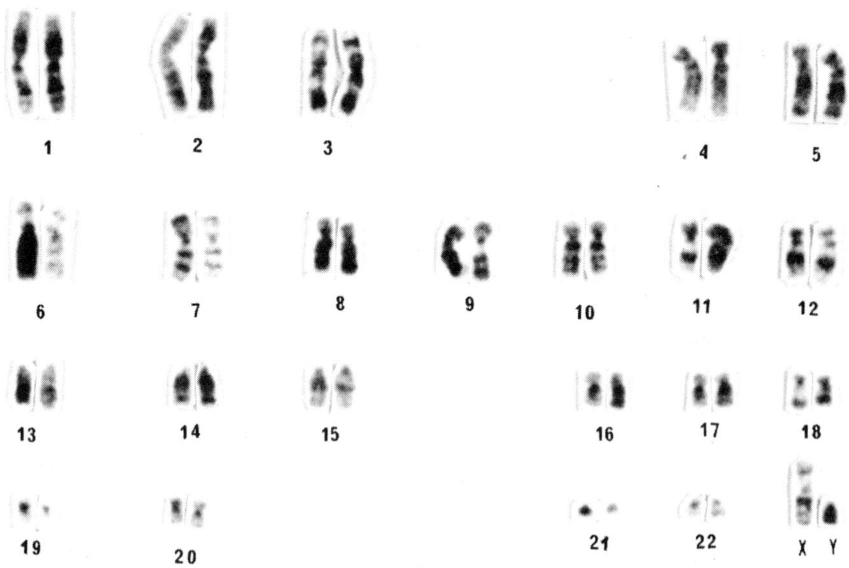

Fig. 2.2 – A karyotype of a human male cell which after spreading on the slide was treated with a dilute solution of trypsin. Note homologue and sex chromosomes.

two X's. The other 44 chromosomes are called *autosomes.* Banded karyotypes can clearly help in determining any loss, gain or rearrangement of parts of chromosomes. Even though a metaphase cell would show 200–400 banded regions, the longer, less condensed chromosomes of prometaphase cells show as many as 2000 bands [1]. Use of prometaphase cells greatly increases the resolution and permits detection of alterations of smaller segments.

The short arm of the chromosome is called *p*, the long arm *q*; the bands are referred to as 1, 2, 3, etc. and are subdivided into sub-groups or sub-bands. Several other techniques are also available for identification of chromosomes (e.g. C-, Q-, R-, T-, and BrdU-banding) [2].

2.4 MEIOTIC CELL DIVISION

Meiotic cell division is responsible for reducing the chromosome number to 23 in sperm and egg cells. Those cells which have been set aside in the gonads as predecessors of gametic populations go through a series of two cell divisions. These *spermatocytes* and *oocytes* replicate DNA and remain in preparatory phases for much longer periods than do mitotic cells. In males this process begins at puberty whereas in females the oocytes complete the preparatory and some further phases while the female embryo is still in the mother's womb.

The essential features of the process of meiosis are as follows. The homologous chromosomes in early prophase pair or *synapse,* bringing the 46 chromosomes into 23 *bivalents.* Soon after, each member of a bivalent becomes recognizable as two chromatids; each bivalent thus consists of a total of four chromatids. Each of the two sets of chromatids is held together at the centromeres. Some exactly equivalent exchanges of segments between non-sister chromatids produce a genetic exchange called *crossing-over.*

The chromosomes condense and the bivalents align on the spindle in metaphase so that the homologous centromeres are on opposite sides of the equator. A set of 23 bivalents with segments exchanged is shown in Fig. 2.3. The alignment of paternal and maternal members of each bivalent during metaphase is random. During anaphase one member of every bivalent moves to the nearer pole. Both sister chromatids move together since the centromere has not yet divided. This separation of bivalents reduces the number of chromosomes to 23 at each pole. The random alignment and migration of bivalents ensures the formation of a remarkably heterogeneous population of gametes.

Completion of the first meiotic division, the reductional division, is followed immediately by the second division with no intervening interphase period of DNA synthesis. The second division is like mitotic division except that there are only 23 chromosomes, each consisting of a pair of sister chromatids. Separation of the sister chromatids ultimately results in four daughter cells each carrying 23 single-stranded chromosomes.

In males four spermatids are produced by meiosis from each spermatocyte. In females, however, meiosis produces only one functional egg from each oocyte. Cytoplasmic division is unequal, so one daughter cell from each meiotic division is a small polar body.

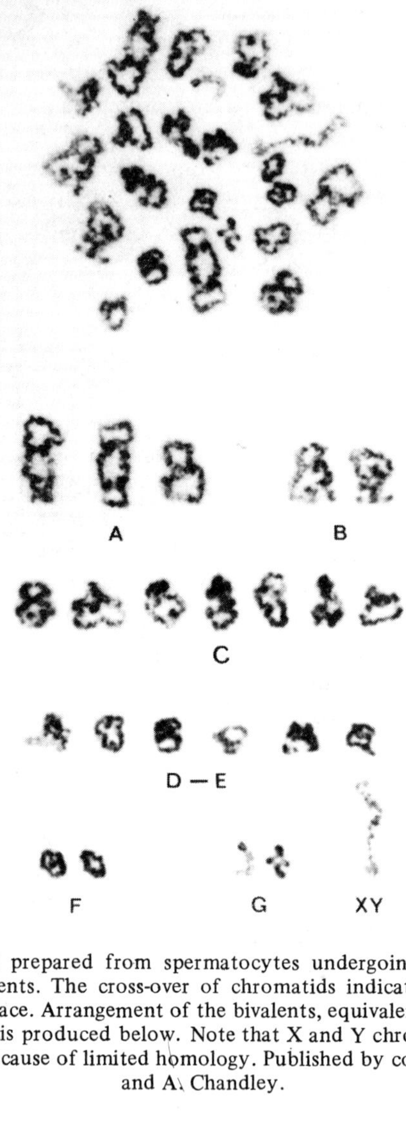

Fig. 2.3 – A cell prepared from spermatocytes undergoing meiotic division. There are 23 bivalents. The cross-over of chromatids indicates that genetic exchange has taken place. Arrangement of the bivalents, equivalent to the karyotype shown in Fig. 2.2, is produced below. Note that X and Y chromosomes pair only in a small region because of limited homology. Published by courtesy of M. Speed and A. Chandley.

2.5 ERRORS OF CHROMOSOME NUMBER IN MAN

In a cell going through the first meiotic division an error of segregation may take place, and then both members of a bivalent migrate to one pole. Similarly, in a mitotic cell or cell in meiosis II both chromatids may migrate to one pole. These errors would result in an extra chromosome in one daughter cell and one too few in the other. If such a cell happens to be the precursor of a sperm or an egg which enters fertilization, the result is a zygote with *trisomy* (three homologues) or *monosomy* (only one chromosome rather than two homologues).

Such mitotic error in a zygote or in one of the cells at a post-zygotic stage may result in a mosaic (i.e. an individual or tissue carrying two or more types of cells).

Such errors of segregation, called non-disjunction, occur rather frequently [3, 4]. Some examples are trisomy 21 (Down's syndrome), trisomy 13 (Patau's), trisomy 18 (Edward's), XO (Turner's), XXY (Klinefelter's), XYY or XXX. Other types also occur but most errors result in prenatal death. Trisomy for chromosome number 16, for example, is the most common type seen in spontaneous abortions but is never seen in live births. Details of these syndromes or photographs of such individuals are not reproduced here because of space limitations. The interested reader may refer to Bond and Chandley [5] or to any textbook of human or medical genetics. Two points, however, must be mentioned: the presence of a Y chromosome results in the male phenotype, and most autosomal aneuploidies result in such genetic imbalance that the embryo cannot survive the prenatal period.

2.6 ABERRATIONS OF CHROMOSOME STRUCTURE

The structural integrity of a chromosome is constantly being challenged by mutagens present in the environment. Defects in DNA repair systems also disrupt the continuity of chromosomes. These result in simple *deletions, duplications* or *inversions* or a segment within a chromosome, or *translocation* of a given segment onto another chromosome. All such events may result in abnormal gametes, depending on the mode of segregation of the bivalent and genetic component of the deficient or duplicate segment. Mitotically, however, structural changes which result in significant loss or gain of chromosome material and/or formation of *dicentrics* (chromosomes with two centromeres instead of one) or *acentric* fragments (no centromere) usually cause death or abnormality of development.

In man, examples of deletions include cri-du-chat (5p), retinoblastoma (13q), and Wilm's tumour (11p). A translocation is usually seen in chronic myelogenous leukaemia where part of the long arm of chromosome 22 is translocated onto the long arm of number 9 (the deleted 22 is called Ph^1 or Philadelphia chromosome). Examples of duplication are many and varied, and phenotypic manifestations depend upon the quantity and quality of genetic material involved.

Some types of structural aberrations seem to result from the genetic constitution of the individual itself. For example, a high frequency of translocations, called quadriradials, along with fragments of chromosomes, are found in patients with Bloom's syndrome (Fig. 2.4). Free fragments are also found in diseases such as Fanconi's anaemia and ataxia telangiectasia. In other instances the cells of certain individuals produce abnormally high incidences of chromosome breakage when challenged by mutagens (e.g. DNA cross-linking agents in Bloom's syndrome, ultraviolet radiation in xeroderma pigmentosum, and X-rays in ataxia telangiectasia). These responses appear to result from faulty DNA repair systems Such genetic instabilities have been discussed recently in some detail [6], as has been the role of chromosomes in human cancer [7, 8].

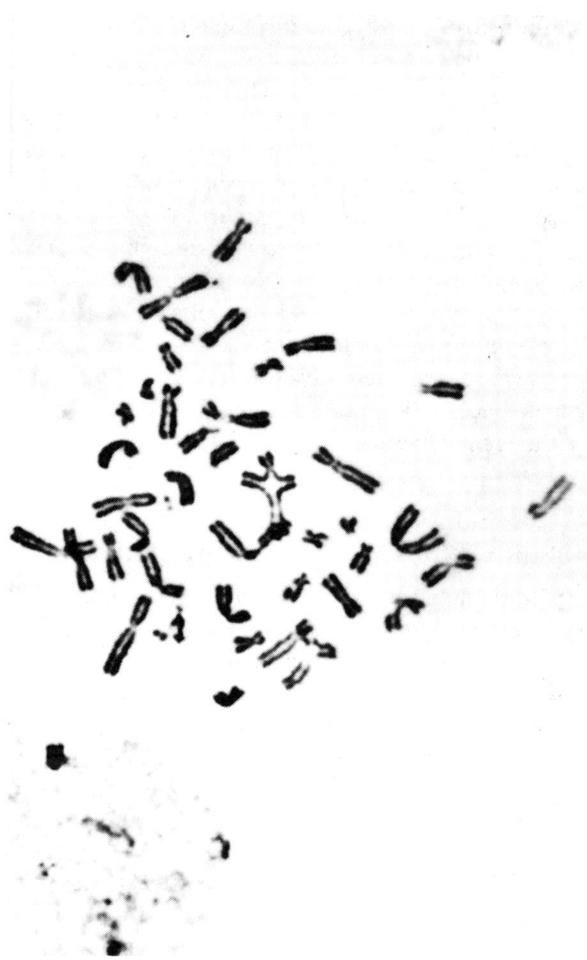

Fig. 2.4 – A mitotic cell from an individual carrying a recessive gene pair for Bloom's syndrome. Chromosome rearrangements, seen as quadriradials, are a characteristic of the disease. Published by courtesy of R. Chaganti.

The frequency of chromosome abnormalities in newborn infants is about 7 per 1000. It is sufficient to state that any imbalance of genetic material, even if minor, would cause developmental aberrations. This is evident from the fact that about 50% of aborted foetuses carry some sort of chromosome defect.

2.7 POINT MUTATION IN MAN

Alterations in DNA structure, such as base substitutions or base additions and deletions constitute the basis for hereditary changes or *mutations*. Depending on the mode of action of a given gene, a mutation may be dominant or recessive to the wild type (most common) allele. Mutations may appear in more than one form, giving a series of multiple alleles. A list of about 3000 mutations known in man is now available [9] and the chromosomal localization for many genes is known [10, 11].

In general, undesirable dominant mutations affect structural components (e.g. bone and cartilage development) and recessive mutations affect genes which have enzymes as products. A few examples of conditions produced by dominant mutations in man are achondroplasia, Huntington's disease, Ehlers-Danlos syndrome, Marfan's syndrome, brachydactyly and polydactyly. Some produced by recessive mutations are phenylketonuria, cystic fibrosis, tyrosinosis, albinism, glycogen storage disease and galactosemia. Some other genes, when in heterozygous form, express incomplete dominance or co-dominance. Examples for autosomal genes are haemoglobin variants (e.g. sickle-cell anaemia, haemoglobin C) and A, B, O blood groups (for AB phenotype only). Mutations on the X chromosome include vitamin D-resistant rickets (dominant) and a number of recessive mutations: colour blindness, haemophilia, hypoxanthine-guanine phosphoribosyl transferase deficiency, and pseudo-hypertrophic muscular dystrophy. With the exception of the gene for H-Y antigen, no clear-cut case of Y-linked genes in man is available.

2.8 COMPLEX PATTERNS OF INHERITANCE

Several characters in man are controlled by genes which, though they follow the Mendelian pattern of transmission, do not function in a simple dominant-recessive fashion. Some are described above as expressing incomplete dominance or co-dominance. Other genes are not expressed in all individuals carrying them; *penetrance* is less than complete. For example, only 80% of all individuals carrying a gene for polydactyly show the character. The same gene may be expressed either fully or to varying degrees of severity. Still other characters are controlled by *polygenic* inheritance in which two or more genes control expression (e.g. skin colour, height and intelligence).

DNA in man, as in other organisms, is also carried in a cytoplasmic organelle, the mitochondrion. This DNA is subject to any and all alterations to which the nuclear DNA is subject. Mitochondrial DNA, for example, expresses racial variations for Hinc II sites ([12]); however no disease condition has yet been found to be definitely associated with mitochondrial genes.

REFERENCES

[1] J. J. Yunis, *Science,* 1976, **191,** 1268.
[2] A. T. Sumner, *Cancer Genet. Cytogenet.,* 1982, **6,** 59.
[3] Th. E. Kelly, *Clinical Genetics and Genetic Counseling,* Year Book Medical Publ., Chicago, 1980.
[4] V. M. Riccardi, *The Genetic Approach to Human Disease,* Oxford University Press, Oxford, 1978.
[5] B. T. Bond and A. C. Chandley, *Aneuploidy,* Oxford University Press, Oxford, 1983.
[6] T. M. Schroeder, *Cytogenet. Cell Genet.,* 1982, **33,** 119.

[7] J. D. Rowley, *Cancer Invest.*, 1983, **1**, 267.
[8] A. A. Sandberg, *The Chromosomes in Human Cancer and Leukemia,* Elsevier North-Holland, Amsterdam, 1980.
[9] V. McKusick, *Mendelian Inheritance in Man,* Johns Hopkin's University, Baltimore, 1978.
[10] V. McKusick, *J. Hered.*, 1980, **71**, 370.
[11] Oslo Conference, Human Gene Mapping 6, *Cytogenetic Cell Genet.*, 1982, **32**, 111.
[12] H. Blanc, Ch. Kuang-Ho, M. D'Amore and C. C. Wallace, *Am. J. Hum. Genet.*, 1983, **35**, 167.

3

Mutagenicity testing – some background information

FRIEDRICH K. ZIMMERMANN,
Institut für Mikrobiologie, Technische Hochschule,
D-6100 Darmstadt, Federal Republic of Germany
and
RHODA E. TAYLOR-MAYER,
Department of Biology, Slippery Rock University,
Slippery Rock, PA 16507 USA

3.1 INTRODUCTION

The basic unit of a living organism is the cell. Some organisms are unicellular, consisting of only one cell, whereas others are multicellular and may consist of astronomical numbers of cells. Multicellular organisms not only consist of many cells but also of many different types of cells, each type performing a specific function at a specific site, at a specific time, and under specific conditions. Even a unicellular organism may change its activities according to environmental conditions. Each organism, unicellular or multicellular, contains a very complex program which directs its activities according to the needs imposed by the environment. This program is stored in the genome and is determined by the sequence of nucleobases in DNA units called genes. Information stored in such genes is used only under well-defined conditions. If required, the information encoded in the base sequence of genes is activated and transcribed into one of three types of RNA. Ribosomal RNA is needed in robosomes and tRNA functions in the translation of the third type of RNA, mRNA, which codes for the amino-acid sequence of proteins. It is imperative for the proper functioning of a cell that the genes are intact and provide the right information which is then correctly transcribed into RNA and from there correctly translated into proteins. It is also imperative in multicellular organisms that the expression of genes be correctly regulated: that only a certain set of genes be activated at a given time and under specific conditions to contribute their information to the transcriptional and translational machinery so that the cell reacts in the proper way at the proper time. Proteins formed at the wrong time in the wrong cell can cause severe adverse effects. The same is true for the failure of a gene to direct the formation of the required gene product.

Even a simple bacterium such as *Escherichia coli* carries about 5000 genes, the proper function of which ensures that this bacterium can thrive in its natural habitat. The genetic program has been adapted through evolution to its present level of efficiency. Any random change in the genetic program will most likely be deleterious; only very rarely will such a random change result in an improved fitness to cope with the environment.

There are still more stringent restrictions for the functional compatibility of random changes in multicellular organisms. The zygote formed by the fusion of gametes divides to form daughter cells which divide again. But very soon diversification begins, into cell types with different functions which realize different and specific subprograms of the total genetic program. This is called *differentiation.* Many cells have to interact in an orderly preprogrammed fashion and this is only possible if they all have the same basic and unchanged program. Consequently, it is not only important that gametes with an intact genetic program fuse to form a zygote with an intact program, but this program has to be transmitted to all the descendant cells. Such descendant cells in multicellular organisms can be divided into *somatic cells* which make up all the different cell types, tissues and organs, and into *germ line cells* which will give rise to gametes for the next generation. From the standpoint of a responsible human being we may conclude that maintaining the genetic integrity of the germ line cells is our responsibility to future generations. On the other hand, from a more selfish point of view, it is important for our own personal comfort to maintain the genetic integrity of our own somatic cells so that they will fulfil their assigned functions correctly. Cancer is caused by cells which have lost control and grow too rapidly and even grow in places where they should not grow. This disobedience is transmitted to their descendant cells and a tumour will form if no efficient mechanisms eliminate such rebel cells.

It took a very long time before biologists began to realize that we may introduce into our environment agents which may damage our genetic program. Muller, in 1927, was the first to demonstrate the induction of mutations (heritable changes) by X-rays in the genetic program of Drosophila [1]. Even though he directly pointed to the dangers of use of X-rays in diagnosis, he was not taken seriously by the medical profession. It was only the immense amounts of radiation released by atomic bomb blasts which alerted the public to genetic damage caused by environmental factors. Today, concern about radioactivity has reached hysterical proportions. In contrast, there seems to be no public concern about the genetic damage caused by ultraviolet light, which geneticists use as a convenient means of inducing mutation in experimental organisms. The medical profession is now very alert to the dangers of too much sunshine, but has not been able to convince the public of this source of health risk.

Induction of mutation by chemical agents was demonstrated about 40 years ago by Auerbach in Edinburgh, Oehlkers in Freiburg and Rapoport in Russia [2]. As time passed, the list of chemical mutagens expanded. Most of those chemicals came from the shelf of the chemist, but quite early, investigations showed that strong mutagens were also used outside chemical laboratories. One example is β-propiolactone, used by surgeons to sterilize their instruments. It was found by Smith and Srb in 1951 to induce mutations in microbes [3]. Not until the mid-sixties did geneticists become aware of the wide distribution of mutagens in the environment. Most of these are products of the chemical industry, and are used as therapeutics, preservatives and pesticides, and in comestics and technical products such as plastics, glues and paints.

The reaction of many people as they realized these hazards of modern society was to go back to nature. However, at the same time it became apparent that there are natural mutagens such as pyrrolizidine alkaloids [4], and mycotoxins in various natural foodstuffs infested with moulds [5].

All this has led to a growing concern about environmental mutagenesis. Our environment, as modified by modern civilization, and even the unmodified environment has to be screened for mutagenic activities and mutagenic pollution.

Mutagenic pollution has received increasing attention as it has become apparent that mutation may be one of the major causes of cancer. This suspicion was voiced early this century by the zoologist Boveri at the University of Würzburg and was later called the mutation theory of cancer by Bauer [6]. In the sixties Druckrey and his associates at the University of Freiburg opened the field of chemical carcinogenesis as they systematically synthesized compounds in order to study their potential carcinogenicity. Such chemicals could then be tested for mutagenicity and a correlation established between the carcinogenic and mutagenic activities of a given agent [7, 8]. At that time, however, the accepted view was that cancer was caused by viruses. The notion that "carcinogens are mutagens" became accepted only after Ames and his colleagues [9] published a list of about 300 carcinogenic chemicals which turned out to be mutagens in a bacterial mutation assay. This correlation test is an expensive undertaking in respect to manpower, installations for animal care and material cost. The much faster and cheaper mutagenicity testing provides for 'short-term carcinogenicity testing'.

Mutagenicity testing is consequently performed not only to identify sources of mutagenic compounds which may induce mutations in germ line cells and thus endanger future generatons, but also to provide an indirect means of identifying carcinogens which may act to induce a variety of genetic alterations in somatic cells, eventually leading to cancer.

There are two major tasks in mutagenicity testing. One is comparatively simple. We must test for mutagenicity any chemical which is newly introduced into our environment as a pesticide, preservative, drug, food additive or for any other use that could lead to exposure of humans to it. More difficult to test are the numerous complex sources which could contain mutagenic components. Testing has to be done not on the pure chemical but on a mixture of an unknown number of mostly unknown chemicals. Different potential sources of mutagenic agents require different approaches, and few standard procedures are available. However, a considerable amount of work has been done and progress has been achieved. It is the aim of this book to provide a few examples of how the urgent problem of screening our environment for sources of mutagenic pollution has been tackled.

In this book we are concerned with detecting and identifying chemicals in our environment which cause genetic damage and cancer in humans. One approach in environmental mutagen pollution analysis would be to use a list of 'known' mutagens and to test environmental samples for their presence by the methods of analytical chemistry. This is an impossible task because of the large numbers involved and also because we do not know all the mutagenic chemicals. The first step is usually to fractionate environment samples and perform mutagenicity tests on the fractions. If a positive response is obtained, the fractions are separated further into chemicals which can either be identified analytically if they are known chemicals, or purified and then have their structure elucidated.

Another approach is to use indicator organisms in the environment to monitor the presence of mutagenic-carcinogenic-teratogenic agents. Two contributions in this volume report on ways to achieve this. Van Kreijl and Sloof (Chapter 6) describes the use of fish to monitor pollution of water bodies, and Plewa (Chapter 10) presents an example of how mutagenic pollution of agricultural land can be monitored with specially developed corn strains. Such biological approaches immediately signal the presence of mutagenic agents in the environment, and the task then is to identify the chemical nature of the mutagenic pollutants.

It may not be warranted to worry about all detectable traces of mutagens in the environment. However, concentration of certain chemicals in the food chain may cause a real genetic hazard to man. We have to live with the fact that our 'natural' environment is full of mutagens/carcinogens. An impressive compilation on this topic and on how our metabolism can cope with such an immense 'mutagenic load' has been provided by Ames [10].

3.2 DEFINITIONS

What is a mutation? The definition is very clear and simple: it is a change in the genetic information which is transmitted through mitotic and/or meiotic nuclear divisions. What we in fact observe, however, is a changed individual, animal, plant or cell which shows different properties. We observe a change in *phenotype,* and the question is whether this reflects a change in the genetic program or *genotype.* The enormous change in pigmentation possible within a given human individual clearly shows that the phenotype can vary without a concomitant change in genotype. Such a variation is called a *phenocopy.* Phenotypic changes which can be transmitted through many cell generations have been reported for the cell cortex structure or surface antigens in the protist Paramecium [11]. Therefore geneticists have had to establish criteria for demonstrating that an apparent change in phenotype is indeed caused by a mutation. In eukaryotes the ideal procedure is to cross the 'mutant' individual with a 'normal' one and check the progeny. A mutation is indicated if the new phenotype follows the transmission pattern of a typical Mendelian factor. A less rigorous test is the demonstration that a given microbial variant generates a large population of descendant cells all showing the same phenotype. Alternatively, in multicellular organisms, a mutation in a somatic cell can generate large patches of tissue expressing a deviant phenotype. At any rate, it is important to be aware of the fact that the appearance of organisms can change dramatically, depending on environmental conditions, without the slightest change in the genotype.

In the old days of genetics, a *mutant* or *mutated allele,* an altered form, was compared with the 'normal' so-called *wild type allele.* However, it has become clear with the refinement of genetic analysis that a given gene can be present in multiple forms with differences in the DNA base sequence even in 'wild' populations not manipulated by geneticists. Whether multiple allelic forms of a given gene co-exist in a population depends on the efficiency of the individuals carrying such alleles for copying with their environment. Most human populations show multiple allelism for many genes (e.g. three different alleles I^A, I^B and I^O are responsible for the ABO blood-group system) and we now say that a mutant allele is any derivative of an original allele. Alleles with only negative effects do not accumulate to a large extent in a population. The term *mutant allele* is a useful term only if we are sure that the allele has arisen as a new form spontaneously or was created intentionally in an experimental organism by chemical or radiational *mutagenesis.* An individual which carries such a newly arisen mutant allele is called a *mutant.*

Eukaryotes carry their genetic information on chromosomes within a membrane-bounded compartment of the cell called the *nucleus.* Additional genetic information is found in certain organelles such as mitochondria and in plastids in green plants. Although additional 'cytoplasmic factors' have been found in eukaryotes, none have been identified in humans. Occasionally foreign genetic elements such as viruses are found in cells. The vast bulk of genetic information, however, is encoded in the DNA present in chromosomes.

Higher eukaryotes carry two copies of each chromosome except for those involved in sex determination, the *sex chromosomes.* Consequently most of the genes are present in duplicate. We speak of a *homozygous* condition if both copies or alleles are identical; a *heterozygous* condition means that the two alleles are different. *Codominance* refers to a situation where both alleles are expressed, as in a person of blood group AB, which has the allelic condition I^A/I^B. Many mutant alleles code for an inactive enzyme, the presence of which is often masked by the 'normal' allele which codes for a functional gene product. A heterozygous carrier can be functionally normal or only slightly affected. This has led to the idea that most mutations affecting only a single gene (usually by changing the sequence of nucleobases in only one position) are *recessive,* whereas the original and functional allele is *dominant.* However, this idea is no longer acceptable. We know a number of dominant genetic defects in humans, i.e. which affect even a heterozygous carrier in the presence of the 'normal' counterpart allele. Moreover, an increasing number of dominant alleles have been obtained in experimental organisms as genetic analysis has progressed.

Mutation can occur spontaneously. That is to say, we do not know why the mutation occurred; the term does not imply anything about the mechanisms involved. The term *spontaneous mutation* only makes sense when seen in contrast to *induced mutation* or *induced mutagenesis* where a deliberately administered mutagenic agent has changed the information of the genome. The fact that spontaneous mutations occur demonstrates that the genetic material is replicated and maintained with limited fidelity. Although mutations occur very rarely, in a very large population of cells (e.g. in a multicellular organism) mutations are constantly occurring. A chemical or physical mutagenic agent enhances the rate of spontaneous mutation.

3.3 TYPES OF MUTATION AND GENETIC CHANGE

As previously defined, a mutation is a hereditable change in the genetic information. Such changes occur in a variety of ways. Elucidation of the genetic code and, more recently, the ability to sequence genes has helped to identify changes at the level of DNA base sequences. These *point mutations* can be classified as *base substitutions,* and *deletions* or *insertions* of a base or short base sequence.

The genetic code consists of 64 possible *codons,* read from mRNA as three nucleotides in sequence, which specify one of the twenty amino-acids found in proteins. There can be up to six different codons for one and the same amino-acid. At the beginning of the coding region of each gene is found the *initiation codon.* Three *termination codons* stop the translation and signal the end of the polypeptide chain.

Base substitution can change one codon into another by substitution of one or more nucleotides. Even though there is a change in the base sequence and a transmissible alteration, such a change may have no phenotypic effect (a *silent base substitution*); for example, one leucine codon may be changed into another leucine codon. However, a base substitution can also change a codon into a codon for a different amino-acid; for example, a change in one nucleotide changes the codon determining the amino-acid glutamic acid into the codon for valine. This is the change responsible for the difference in the β-chain of haemoglobin which causes sickle cell anaemia. A simple base substitution has been reported also for the Ferrara form of β^{o} thalessaemia, where the codon for glutamine has been changed into a termination codon. This change leads to a premature chain termination during translation of the haemoglobin β-chain, resulting in a non-functional polypeptide fragment [12]. Another base substitution with a grave effect was reported by Spritz *et al.* [13], also involving the haemoglobin β-chain. This base substitution, however, did not affect a coding region but affected one of the non-coding *intron* regions interspersed in this gene. Such introns are transcribed into mRNA but must be excised before such an mRNA molecule can be translated into the amino-acid sequence of a functional protein. The single nucleotide change in the intron region prevented excision of the intron so that non-functional β-chain was synthesized. These examples illustrate the fact that even single base substitutions can have drastic effects.

Proper translation of the genetic informaton encoded by mRNA requires that the genetic message be read in the correct reading frame. Mutations based on the insertion or deletion of one base, or of some number of base pairs which cannot be divided by three, results in a shift of the reading frame. Such *frameshift* mutations have been observed in the Wayne variant of the human alpha chain. Codon 139 of the Wayne mutant allele has been shortened by the deletion of one nucleotide. The consequence is that the remaining two nucleotides are read together with the first nucleotide of the next codon; the second two bases of that codon are read with the first of the third codon nucleotides, etc. Normally the fourth codon should have signified termination, but shifting the frame of reading means that the second two bases of the fourth codon are now coupled with the first base of the next codon to form a codon which signifies an amino-acid. Since there is no longer a termination codon, translation continues, resulting in the synthesis of an extra long alpha chain.

Chromosomal numerical aberrations are the most drastic change in a genotype. There are two classes of numerical aberrations: *ploidy mutations* and *aneuploidy.* Ploidy mutations lead to the duplication of complete sets of chromosomes. Higher organisms are diploid; they have two sets of chromosomes. Mutations are known which lead to three sets (*triploid*) or four sets (*tetraploid*). Aneuploidy refers to conditions where the chromosome numbers differ from the regular number of a complete set of chromosomes or multiples thereof. Maintenance of the normal haploid or diploid chromosome number in mitosis and meiosis requires the proper function of the *spindle fibre apparatus* which moves chromosomes to the opposite poles of the spindle. The fibres are *micro-tubules* of *tubulin* and associated proteins. Chemicals are known which interact specifically with tubulin. The best known such chemical, colchicine, has been used in plant breeding to induce polyploid strains by inhibiting mitotic chromosome movement. Cytologists use it for the same purpose. Such chemicals need not induce any other type of mutation; thus they will be regarded as non-mutagenic in most of the popular tests.

Changes in chromosome number leading to *aneuploidy* are relatively frequent in human populations. Three syndromes are frequently seen: trisomy 21, or Down's syndrome, occurs once in 600 live births; Turner's syndrome, based on the presence of only one X chromosome, XO, rather than the normal female condition XX, occurs in about 1 in 10000 live births; Klinefelter's syndrome, XXY, is found in about 1 in 1000 male births. Occasionally individuals showing these syndromes are mosaics. They contain both normal and aneuploid cells which suggests that chromosomal mal-segregation occurred in mitotically dividing somatic cells rather than in the gametes which produced the individual at fertilization [14].

Structural chromosomal aberrations can easily be detected in cytological preparations and have been shown to exist in humans. Chronic myelogenous leukaemia is associated with a chromosomal re-arrangement. Part of chromosome 22 is translocated, usually to the end of chromosome 9. Several other disorders have been shown to be correlated with chromosomal structural aberrations, usually deletions of part of a chromosome. Translocations where segments of one chromosome are transferred to another different chromosome can be the cause of low fertility. This also applies to inversions where a segment of a chromosome has been inverted relative to adjacent parts. Consequences are particularly grave when the inverted segment includes the centromere.

Like all other organisms, humans can be invaded by viruses. A virus carries its own genetic material in the form of either RNA or DNA. In a compatible cell viral nucleic acids can be genetically active, i.e. they can direct cellular metabolism according to their own program. Some viruses kill cells, and others can persist for a long time and be transmitted to subsequent cell generations. It is now well known that tumour viruses can be integrated into host chromosomes, at least in experimental animals [15], and thus become an established part of the host genome, being replicated and transmitted like normal genes. This is fully equivalent to a mutation: a stably transmitted alteration of the genome, resulting from the integration of foreign genetic material. Such viral infections have been associated with cancer in a number of experimental animals; this should reconcile the controversy between the mutation and viral theories of cancer.

Bacterial viruses can also be integrated into host DNA, where they are replicated and transmitted. Occassionally they are released from the genome and multiply like typical viruses, killing the host cell. This release can be precipitated by treatment with mutagens. This process, called *phage induction,* has been used as a mutagenicity test [16].

Heterozygosity in the genotype of a diploid cell or organism provides the basis for an additional type of genetic variation not involving *de novo* mutation. Since a given locus is represented by two different alleles, the effects of such a heterozygous condition on the phenotype depend on how the two alleles interact. A considerable number of genetically determined metabolic disorders are not expressed in the heterozygote because the defective allele is present with a normal and functional allele. A single copy of the fully functional allele is usually sufficient to support the required function and is therefore called *dominant.* Any genetic event leading to elimination of the functional allele will result in the disease: loss of an entire chromosome, loss of a chromosomal fragment, or point mutation of the functional allele. These events might be considered to be mutations, but they are not the only possible mechanisms which lead to the expression of recessive alleles.

Stern [17] observed mosaic spots in adult Drosophilae which were heterozygous for two recessive alleles. One chromosome carried a recessive allele causing a pigment defect in the absence of a functional allelic partner. The other homologous chromosome carried another recessive allele which, when expressed, leads to bristles looking *singed.* Many such doubly heterozygous flies looked normal, but some showed spots with a pigment defect, others had singed bristles. However, a few flies exhibited really baffling aberrations consisting of pairs of spots of equal size and shape, one of which expressed the pigment defect, the other showing singed bristles. Stern interpreted this twin-spot phenomenon as a result of *mitotic recombination.* It is generally accepted that recombination between homologous chromosomes is a standard process during meiosis. It is less well known that a much rarer event, mitotic recombination, occurs. Mitotic recombination can have drastic effects in individuals heterozygous at many gene loci. A single mitotic crossing-over can lead to expression of all the markers located on the far side – relative to the centromere – of the crossing-over event. Mitotic recombination, especially mitotic crossing-over, has been demonstrated since in fungi, other insects, green plants and also in mammals [18, 19].

Consequently, mitotic recombination has to be considered as an additional mechanism of genetic change in mitotically dividing cells of heterozygous diploid organisms.

Mitotic recombination can be induced by mutagens. Mutagenicity testing for mitotic recombination has been developed with fungi such as *Aspergillus* [20], the yeast *Saccharomyces cerevisiae* [21], green plants [22], and finally mice [19].

It has to be re-emphasized at the end of this section that genetic change is relevant not only for germ line cells. The proper function of multicellular organism requires that each cell play 'by the rules', i.e. obey all the control signals which coordinate the many different functions in all different cell types.

3.4 MUTAGENIC CHANGE AND REPAIR SYSTEMS

As previously stated, the term spontaneous mutation implies that we do not know the precise cause of an observed mutation. Spontaneous mutations are basically rare and some types are rarer than others. Chromosomal mal-segregation in human meiosis reaches several per cent [14], but point mutations are estimated to be as low as 1 substitution per 10^7 base pairs. Accurate determinations are difficult, especially in the rarer categories, because we do not usually count aneuploid cells or DNA molecules with substituted base pairs. We normally count individuals with a changed phenotype, and from that we try to estimate the rate at which certain types of mutations occur. The problem is that it is difficult to detect all changes in phenotypes, and not all changes in phenotype are caused by mutation – they can also be caused by uncontrolled changes in the environment.

Yet mutations do occur. Mutation can be truly spontaneous because the fidelity of DNA replication, the accuracy of maintenance of DNA, and chromosomal integrity as well as the precision of chromosomal segregation in mitotic and meiotic nuclear divisions, are limited. The demands for precision, at least at the level of DNA replication and maintenance, are enormous. Let us assume that a bacterium has a set of 5000 genes coding for that many proteins, each with an average length of 333 amino-acids. This requires 1000 coding base pairs per protein and 5×10^6 coding base pairs per genome. We can ignore at present the non-coding regions involved in the control of transcription and translation. A replication error rate of 1 per 10^5 base pairs would appear to be better than any human operational error rate. However, in the case of the bacterium, this would mean that each round of replication would result in an average of 50 'mutations'. Some of the base substitutions could be silent; let us assume 50% of them. This would still leave the new genome with 25 amino-acid replacements, an error rate which would mean that no generation could produce identical replicates. Much higher accuracy, however, is demanded and obviously achieved. Proof-reading activity after DNA replication eliminates replication errors. This is achieved because the template strands are already modified by methylation of certain bases before the onset of replication. This directs the proof-reading machinery, after replication, to selectively remove the 'wrong' base from the newly replicated strand at a site of base-pairing mismatch [23]. However, errors besides those in replication have to be corrected. For example, cytosine can undergo spontaneous deamination to give uracil. The uracil is removed by a specific uracil-DNA-glycosylase and then replaced in a lengthy 'repair operation' by cytosine [24]. Methylcytosine, a modified form of cytosine, can also be deaminated spontaneously and converted into thymine. In the case of thymine originating from methylcytosine, there is no specific removal of the wrong base. Only general *mismatch repair* can remove this irregularity from DNA by removing either the thymine part, which would be correct, or the guanine part of the mismatch, which would then lead to a change from a cytosine/guanine to a thymine/adenine base pair. Indeed, certain sites containing methylcytosine are unusually mutable: '*hot spots of mutation*' [25].

Other types of chemical changes and even breaks in DNA molecules occur spontaneously as a consequence of the limited thermodynamic stability of the DNA double helix, and cellular repair activities are provided for these as well.

In addition to proof-reading and DNA maintenance, there are also repair activities directed against damage induced by mutagenic agents. The existence of several of these repair activities can be demonstrated by mutant alleles which increase sensitivity to the lethal effects of mutagens and also by an increased spontaneous mutation rate, so-called *mutator mutations*.

Ultraviolet light of wavelength about 260 nm induces, among other changes, the formation of dimers of adjacent pyrimidine residues. Many organisms have a so-called *photoreactivating system* which splits the dimers into monomers in the presence of light of longer wavelengths. The so-called *dark repair* consists of a series in which an endonuclease recognizes a pyrimidine dimer and cuts next to it into the DNA strand. This results in a single-strand break which is enlarged to a single-strand gap through the action of an additional nuclease. The gap is filled by *repair replication* until the discontinuity is left only on one side of the newly synthesized tract of DNA. A ligase links this to the 'old' DNA on the other side of the gap. Such a cycle is called *excision repair* and it works as a system of *error-free repair*. Mutants defective in excision repair have been isolated in bacteria and in fungi. Xeroderma pigmentosum, a human genetic disorder leading to light-induced skin lesions, is caused by defects in such a repair system. Mutants with defects in excision or error-free repair are often subject to unusually high mutation rates.

The cellular systems involved in repair of damage induced by ultraviolet light have been investigated with the aid of mutants showing very high sensitivity to ultraviolet light. Some such mutants have a truly remarkable property: ultraviolet light will not induce point mutations in them. Apparently, point mutations obtained after irradiation with ultraviolet light are not a direct effect of the radiation, but a consequence of an *error-prone repair* system. Such mutants have been obtained in bacteria and also in yeast, where the mutation *rad6* eliminates induction of point mutation by a wide spectrum of mutagens

[26, 27]. The important message for the investigator engaged in mutagenicity testing is that a point mutation, e.g. reversion of an isoleucine-requiring yeast cell to an isoleucine-independent reverse mutant, may be the end-product of a series of reactions and not necessarily the primary effect of some agent on DNA.

There are at least three different major repair pathways involved in eliminating genetic damage induced by ultraviolet light: excision, error-prone, and double-strand break repair pathways. When all pathways are operative the repair capacity is tremendous. About 27000 pyrimidine dimers per genome have to be induced in a culture of normal haploid yeast cells in order to kill 63% of the cells. However, when all three repair pathways are blocked by mutations, as few as only 1 or 2 pyrimidine dimers can achieve the same effect [28].

It is not clear at present how many genes are involved in ensuring the accuracy of DNA replication, DNA maintenance and repair of mutagen-induced damage. Haynes and Kunz [27] list about 140 genes in the simple eukaryote yeast which can be mutated to alleles which cause increased sensitivity to mutagens or an increase in spontaneous mutation.

The differences between a fully repair-competent microbial strain and mutant derivatives with defective repair systems have been exploited in mutagenicity testing with the so-called repair tests. The basic concept is that a chemical inducing repairable damage in DNA will be more toxic and lethal to the repair-deficient mutant than to the fully repair-competent original strain.

3.5 CHANGES IN THE PATTERN OF GENE EXPRESSION

Studying genetic effects in mammals or in mammalian tissue culture cells is complicated by the differentiated state of almost all cell types. In a specialized cell only a limited number of genes will be active, the majority being 'silent'. Holliday and Pugh [29] proposed that methylation of cytosine residues could play a role in differentiation, and evidence has accumulated recently which indicates that 'silencing' of genes in mammalian cells can be achieved by the methylation of cytosine residues in specific groups of nucleobases. 5-Methylcytosine (abbreviation C5m) is the most common modified base in mammals and is usually found in 5′-CCGG-3′ or even 5′-CG-3′ stretches in DNA. Methylated DNA sequences are replicated normally. The methylated strand serves not only as the template for DNA replication but also directs the methylation pattern. A 5′-C4mCGG-3′ sequence directs the formation of a 3′-GGCC-5′ sequence:

5′-C5mCGG-3′
3′-G GCC-5′

in a double helical stretch of DNA within the replicated molecule. The central 5mC base directs the methylation of the central cytosine residue in the opposite strand:

5′-C5mC G G-3′
3′ G G5mCC-5′.

Consequently, once established, the pattern of methylated cytosine residues can be stably transmitted, provided the methylating enzyme *S*-adenosyl-L-methionine DNA (cytosine-5) methyl transferase is acting properly and is supplied with *S*-adenosyl-L-methionine [30].

Systems of '*de novo*' methylation also exist. DNA from certain viruses can be integrated into the chromosomes of mammals, where they are replicated. The DNA of the viral particle is usually unmethylated, but is methylated when integrated into the host chromosomal genome. Methylation in the controlling regions of viral genes is correlated with 'silencing' of such genes [31, 32]. The general conclusion is that methylation of specific bases in specific base sequences plays a role in the regulation of gene expression but may not be the only mechanism of regulation.

Agents and conditions which interfere with methylation will lead to 'undermethylated' DNA, which can change the physiological properties of mammalian cell lines so that they express genes which they would not normally express. These changes are quite stable; clones can be established, and the stability could be considered as the result of a true mutation.

A potent inhibitor of methylation in general is L-ethionine. Christman *et al.* [33] treated Friend erythroleukaemia tissue culture cells with L-ethionine and observed the formation of α- and β-globin mRNA, haeme and a number of erythrocyte-associated enzymes and at the same time could demonstrate that the newly formed DNA was 'hypomethylated'.

In 5-azacytidine, an analogue of cytidine, the carbon atom at position 5 in cytosine is replaced by a nitrogen atom, so that methylation is no longer possible. This analogue can be incorporated into RNA and DNA by the various enzymatic reactions of the so-called pyrimidine salvage pathway [34]. The cytosine-methylating enzyme is probably inactivated when it encounters a 5-azacytidine residue and this results in a reduction of methylating capacity and undermethylation of DNA [33], with associated expression of 'silent' genes. Constantinides *et al.* [35] found, for example, that 5-azacytidine could convert cells in tissue culture, which were not muscle cell precursors, into twitching muscle cells.

Mammalian females carry two X chromosomes. A random 'inactivation' of one or the other X chromosome occurs during early embryonic development, so females heterozygous for genes on the X chromosome are mosaics. Some patches of tissue express one allele, different patches the other. 5-Azacytidine can activate 'silenced' genes on the X chromosome, as shown for the hypoxanthine guanine phosphoribosyl transferase (HPRT) locus in humans [36].

3.6 MUTATION vs EPIGENETIC EFFECTS

Our genetic thinking is still very much influenced by microbial models. Untimely expression of a gene function has been obtained in bacteria and fungi and in all cases could be demonstrated to be the consequence of a mutation. For example, yeast cells change their physiological activities according to the environmental conditions. They form the enzyme maltase only when maltose is present in the medium. However, clones can be obtained which form maltase in a medium without maltose but containing glucose. This variation can be shown to be due to a regulatory mutation which segregates in crosses like a typical Mendelian factor.

The previously described activation of 'silenced' genes, such as becoming a twitching muscle cell or a Friend leukaemia cell producing haemoglobin and other proteins typical of an erythrocyte, could be considered to result from a regulatory mutation, by analogy to the microbial models. However, such altered clones are not due to a mutational change of the base sequence of a gene but are most likely caused by a change in the methylation pattern. This is an important difference. A change in the genetic information caused by a change in the base sequence is not readily reversible. A change in the methylation pattern leaves the genetic information intact and the change can be reversed. There is a change in expression of the genetic information but not in the kind of information present. Therefore we have both mutational and *epigenetic* changes.

An interesting case of reversible 'genetic' change has been reported by Harris [37], for Chinese hamster cells. 5-Bromodeoxyuridine is an analogue of thymidine and is incorporated into DNA by a salvage pathway initiated by thymidine kinase. Cells with DNA rich in 5-bromouracil are very sensitive to ultraviolet light, and resistant clones can be selected. Since resistance is caused by a reduction in thymidine kinase activity, extended selection finally results in cells completely lacking the enzyme. Revertants which have a restored ability to make a functional thymidine kinase can be selected on a special HAT medium. Such revertants occur very rarely, like mutations. However, up to 0.3% conversion into revertants could be obtained when cells were cultured in the presence of 5-azacytidine and L-ethionine, two agents interfering with DNA methylation. Another chemical, sodium butyrate, also led to a 'massive conversion' from thymidine kinase deficiency [37]. L-Ethionine is certainly not a powerful mutagen and has been found to be non-mutagenic in many assays [38]. In many studies of its effects on the activation of silent genes 5-azacytidine was found not to be mutagenic. It was mutagenic in yeast, but only at concentrations 1000 times higher than those used in the Chinese hamster experiment [39]. In contrast, the standard mutagen ethyl methanesulphonate, EMS, was much less active than the other three 'non-mutagenic' agents restoring thymidine kinase proficiency [37]. Apparently the reduction of thymidine kinase in the resistant clones was not caused by a mutation in the base sequence, but by an epigenetic increase in the methylation pattern. This increases in methylation was then reversed by 5-azacytidine and L-ethionine.

Wilson and Jones [40] used DNA with one methylated and one complementary unmethylated strand as a substrate in a methylation assay. The DNA used as substrate had previously been exposed to a number of mutagens: benzo-(*a*)pyrene, 9-aminoacridine, methyl-*N*-nitro-*N*-nitrosoguanidine (an extremely potent mutagen and carcinogen), EMS, ethyl nitroso-urea, methyl nitroso-urea and other alkylating agents. They all inhibited methylation of the unmethylated DNA, which is not surprising since these mutagens induce changes in DNA and thus may interfere with the interaction of the transmethylase enzyme with its DNA substrate.

As mentioned previously, one of the reasons for the great interest in mutation assays is that they can be used as rapid carcinogenicity tests since the correlation between mutagenicity and carcinogenicity of chemicals is quite strong. This correlation supported the mutation theory of cancer. Holliday [41] added in "A new theory of carcinogenesis" that induced changes in the methylation pattern of DNA may also result in regulatory disorders and cancer. Such changes in methylation can occur by interference with (1) the activity of base-modifying enzymes such as transmethylases, (2) the formation of their substrates such as *S*-adenosyl-L-methionine, or (3) DNA, the target molecule of modification.

Clearly there can be two consequences of chemically induced effects in DNA. The classic type of mutation results when a primary lesion of DNA is converted into a 'new' base sequence in the course of an error-prone, inaccurate, repair process. Not all types of induced alterations of DNA result in mutation, as we have seen. Some may only interfere with the systems of DNA modification, such as methylation, and thus cause only epigenetic effects. A correlation between the mutational and epigenetic effects of agents could be the basis for the empirically established correlation between the mutagenic and carcinogenic activities of many chemicals.

Mutation and epigenetic effects have DNA as a common primary target. Chemical agents may affect other targets. An agent interfering with an enzyme involved in an error-free repair pathway may cause mutation, but not induce epigenetic effects. In contrast, L-ethionine, inhibiting a DNA methyltransferase, can cause drastic epigenetic effects but may not be mutagenic.

We cannot predict with any certainty how frequently agents can be epigenetically active but non-mutagenic, or mutagenic but epigenetically inactive. We can assume, however, that an agent which induces chemical alterations in DNA most likely causes both types of effects, although with different efficiencies. Agents interfering with the enzymatic machinery of DNA replication, repair or maintenance are likely to be mutagenic; agents interfering with the enzymes involved in epigenetically relevant DNA modifications are likely to cause only epigenetic effects. Interference of a chemical with the activity of a given enzyme depends on highly specific structural parameters. In contrast, any reactive agent can alter DNA structure and enzyme structure, although there is a distinct difference between the two targets. The section of DNA representing a given gene in a diploid cell is present in two copies before, and four copies after, DNA replication. Subsequent nuclear division distributes the replicated chromosomes identically in the daughter nuclei. In comparison, thousands of molecules of a given enzyme are usually present in a cell. A chemical interfering with one or two or at most four copies of a given gene will cause detectable effects, whereas alteration in even 100 molecules of an enzyme present in 5000 copies may be irrelevant. Consequently, a gene is a highly vulnerable target whereas a given enzyme may not be, because it is present in such large numbers.

Some chemicals, however, may block enzyme activity although they do not change the structure of the enzyme. Certain substrate analogues or analogues of allosteric effectors regulating enzyme activity can quite selectively, but reversibly, inhibit the function of all molecules of a specific enzyme. All that is required is a critical concentration of the agent. This concentration will block function in all cells simultaneously. When a typical mutagen reacts with DNA, the average number of 'hits' will be more or less the same in all cells and will be randomly distributed over the total genome of each cell. Since an individual gene represents only a very small portion of the genome, an individual gene will rarely be hit; consequently induction of mutation in this particular gene will be extremely rare. Mutation frequencies in a bona fide mutation assay are very low indeed. In the case of 5-azacytidine and L-ethionine, as reported above, massive 'reverse mutations' of thymidine kinase-deficient mutant cells to a thymidine kinase-positive phenotype were observed. Such massive effects indicate that the observed changes may not be due to mutation but rather reflect some epigenetic condition. The novice investigator will be delighted to observe such massive effects; the experienced geneticist will be quite uncomfortable with unusually high mutation frequencies. Geneticists working on microbes have always marvelled over the high mutation frequencies obtained in some of the mammalian tissue-culture cell systems. Thier scepticism was well warranted. At least in some cases, the effects were not caused by mutation but by interference with transmethylase activity required for the maintenance of a 'differentiated' state. This explains also why some of the potent 'mutagens' are less effective in these systems than are L-ethionine or sodium butyrate, agents which are non-mutagenic in almost all established and genetically defined mutagenesis tests.

3.7 PROBLEMS IN SELECTING A MUTAGENICITY TEST BATTERY

The goal of mutagenicity testing and of the efforts discussed in this volume is to identify chemicals which cause genetic damage in humans. There is no easy solution and the field is fraught with problems. It is important to use test systems in which the changes observed are genetically defined and, above all, not mimicked by effects unrelated to genotypic change. Most mutation tests are based on the detection of phenotypic changes; very few are designed to detect directly changes of chromosome structure or chemical-physical alterations in DNA. The advantage of the recessive lethal test on Drosophila is that the effects observed are indeed changes in genes and these changes are transmitted as regular genetic factors through subsequent generations. A microbial reverse mutation assay measures the restoration of an originally defective allele to a functional condition through a mutational event. This cannot be detected directly, but the consequence of this genotypic change, which allows the cell to grow on a medium selective for the restored function, can be observed as the formation of a colony containing millions of cells. The genetic basis of this can be verified in a few representative samples by performing genetic crosses.

In contrast to this, microscopic observation of broken chromosomes as a result of a treatment seems to provide the most direct evidence of mutagenic action. However, we cannot be absolutely sure that the broken chromosomes are transmitted to the next cell generation. Damage of the genetic material which simply leads to a dead cell is not genetically relevant. Even though a cytogenetic test can impressively demonstrate direct damage to the genetic material, it does not automatically prove the induction of transmissible genetic damage and therefore mutation.

There are also tests designed to detect chemical modifications of DNA bases or breaks in DNA strands. Although these directly study effects on the genetic material DNA, the tests do not prove that the effect is equivalent to mutation. Even though, as we have discussed, chemical or physical modifications of DNA do not automatically result in mutation, such tests have been calibrated or validated by using many mutagens and non-mutagens to show that the observed effects really are specific for mutagens.

There are simple short-term tests which have well understood genetic bases, have been widely used for mutagenicity testing, and are well validated [42] and their strengths and weaknesses are known. They are fast, cheap and do not require expensive equipment. Their disadvantage is that they do not prove that a chemical indicated by the test to be mutagenic is mutagenic to man as well. We would like to identify agents which are 'human mutagens'. We cannot do this experimentally except by use of human tissue-culture cells and there we do not have the full potential of human activation nor the full spectrum of genetic changes. Consequently, we use other organisms and then identify 'mammalian', 'Drosophila', 'yeast' or 'bacterial' mutagens. Considering the fact that 'DNA is DNA' we can be relatively certain that many of those mutagens are 'universal mutagens', but we do not know the actual risk to humans. A mammalian specific locus test would be ideal and a two-year carcinogenicity test in three mammalian species may satisfy critics, but at what cost! Consequently we have to content ourselves with a less perfect approach and use whatever we can afford. Practicality is important and the task to be solved truly monumental.

Practicality is certainly not only an academic aspect. It is no problem to get enough of a pesticide to conduct long-term carcinogenicity tests, but we want to know about the presence of mutagenic pollution not only in prepared food or drinking water, but also in other environmental areas (such as air) where collecting large samples with sufficient material could be difficult. A few milligrams are enough for a microbial short-term test, but gram quantities may be required for a full-scale carcinogenicity test. Time may also be an important factor. Some problems are urgent and it may make a difference whether a short-term test identifies a new source of mutagenic pollution within two weeks or a full-scale mammalian carcinogenicity test identifies an agent only after two or three years.

A survey of presently used mutagenicity tests and their application has been presented in the *Handbook of Mutagenicity Test Procedures* [43]. This book does not evaluate the merits of the different tests and so far no authors have dared commit themselves on such evaluation. Persons engaging in a mutagenicity testing project have to make a final decision on choice of test(s), and such decisions have been repeatedly based on a compromise between the cost of a test, its use in other programmes and the closeness of its approximation tests on humans. At present, it is common practice to use a microbial short-term test and some mammalian tissue-culture assay. Tests *in vivo* with whole animals are used for more specific purposes of detailed quantitative risk assessment. The first step in mutagenicity testing in environmental pollution control is simply to identify sources of mutagenic/carcinogenic agents.

A problem arises in establishing a test battery which covers the entire spectrum of genetic alterations that can occur in humans. For example, Vogel and Natarajan [44] compared the induction in Drosophila of recessive lethal mutations and translocations by a number of mutagens. Many of the agents were potent inducers of recessive lethal mutations but barely induced translocations. This finding demonstrates that it is not sufficient to use a single convenient assay unless it has been demonstrated that that assay reflects the genetic activities of a wide spectrum of mutagens with differing primary effects.

The popularity of the Salmonella reverse mutation assay has made this the first choice in almost all efforts to screen for mutagenic pollution. This is certainly inadequate, and such a test should be supplemented by further tests. Adding more tests to a screening programme causes additional expense for investment in equipment, consumable supplies, personnel and testing experience. In the long run a balance has to be struck between the cost of a 'complete' test battery and the potential increase in information. What can be gained by supplementing the economical and well calibrated Salmonella test with one or two more tests? Nestmann reports in this book (Chapter 8) on the results obtained by adding yeast point mutation and mitotic recombination assays to a programme designed to screen paper-mill effluents for mutagenic pollution. In this particular case the addition of two tests led to the detection of considerably more mutagenic pollutants, but it is difficult to predict in general the performance of enlarged battery of several test systems, compared with that of a single microbial assay such as the Salmonella reverse mutation test or a yeast mitotic gene conversion assay. Even these tests have been made increasingly sensitive. Elimination of error-free repair so that induced DNA damage is repaired by the error-prone pathway has been used to make the Salmonella test more sensitive to low levels of mutagens. The Salmonella reverse mutation test also employs several strains, since no single mutant allele can be induced to revert with all mutagens.

In an international collaborative programme 42 coded chemicals, carcinogens and non-carcinogens, were distributed to investigators known to be using a wide variety of mutation assays [38]. This programme was not only a test for the performance of different assay systems but also for the performance of the investigators. Sucrose was found to be mutagenic in three tests, ascorbic acid in ten, L-methionine in eight. Twelve different groups used a standard Salmonella reverse mutation assay. Eight chemicals were found to be mutagenic by all participants; four agents were uniformly judged to be negative. This means that for less than one third of the chemicals, 12 out of 42, the same test result was obtained in all participating laboratories. The most remarkable result was that no chemical was uniformly negative or uniformly positive in all assays used (including repair, bacterial mutation, and various eukaryotic *in vitro* and *in vivo* systems). A possible conclusion is that everything will eventually be found to be mutagenic if enough persons perform enough different mutagenicity tests. A more positive conclusion is that mutagenicity assays are complex biological tests which have to be executed carefully and evaluated judiciously.

3.8 METABOLIC ACTIVATION

Finding the proper genetic tests is only one of the problems in mutagenicity testing. The second major problem is that many mutagens and carcinogens are not directly mutagenic or carcinogenic. They are *premutagens* and *precarcinogens.* The active forms, the *ultimate mutagens* and *ultimate carcinogens,* are formed during metabolism in certain tissues in certain organisms. The prime organ for metabolic conversion in animals is the liver; this prompted Malling [45] to use extracts free from liver cells to activate premutagens in a fungal mutation test. This seminal discovery became the basis of the liver-microsome Salmonella reverse mutation test developed by Ames *et al.* [9]. A large number of carcinogens have been identified as mutagenic by using this approach. However, there are numerous shortcomings: (1) not all metabolic activation occurs in the liver, (2) the activation capacities of the liver depend on the diet and the induction of certain metabolising enzymes, and (3) there are species differences, and within a species there can be differences between strains and even between sexes. As with the choice of the 'right' mutation test, we also want to obtain the 'right' answer in our metabolic activation systems. Although most chemicals activated in a rodent liver will be activated in a human liver, some will not and *vice versa.*

An interesting approach to harness the metabolic activation capacity of different organs of the mammalian body has been developed by Mohn and Ellenberger [46]. Microbial indicator organisms are injected into the tail vein of rodents in this *intrasanguineous host-mediated assay.* The microbes are distributed throughout the entire body and can be recovered and tested for induced genetic change. Of course, this is more laborious than a simple plate test with microsomal preparations added to the indicator cells.

The *predictive validity* of a given mutation test or mutation test battery depends not only on the mutations induced, but also on the activation system used. We stand a low chance of making a 100% correct prediction; we have to keep this in mind. In spite of all the shortcomings, however, a number of dangerous chemicals have been identified by using relatively simple mutation assays, and a moderate optimism is warranted.

Metabolic activation of premutagens and precarcinogens can occur anywhere in the biosphere. The metabolic activation potential of crop plants has been largely ignored. The contribution of Plewa in this book (Chapter 10) reports on the development of systems to include the metabolic activation of plants in environmental mutagenicity testing. The food chain must also be considered since some chemicals are concentrated in certain organisms and can be activated there. This concentration and activation is the topic of the contributions of Parry (Chapter 7) and of Van Kreijl and colleagues in this volume. Such approaches are indispensable in testing the environment for mutagenic/carcinogenic pollution.

Since metabolism consists of activating chemicals so that they can eventually release energy in APT or similar products useful for cellular metabolic activities, each reactive intermediate is a potential mutagen. It could react with DNA, DNA precursors, enzymes of DNA metabolism, components of the systems for macromolecular synthesis, structural non-DNA components of chromosomes, or the components of the spindle fibre apparatus. It is conceivable that in the long run almost every chemical may be found to be mutagenic since all organic chemicals are metabolized somewhere in the biosphere. Consequently, we have to consider sources of metabolic activation in respect to their relevance to humans.

3.9 THE PROBLEM OF DETOXIFICATION

Mutagenic chemicals may be missed because they have not been adequately activated. In addition, mutagenic agents may be missed because of detoxification in the test organism. Cells not only have systems for the repair of genetic damage, they also have enzymatic systems or special molecules engaged in detoxifying dangerous chemicals. A simple organism such as a fungus may not efficiently detoxify an agent such as dichlorvos, which is both a classic microbial mutagen and a classic false positive in respect to higher eukaryotes. Obviously, the agent must be detoxified [47]. However, a chemical which is very mutagenic in a wide spectrum of microbes may still present a genetic hazard to humans after mammalian tests have shown it to be non-mutagenic or non-carcinogenic. Widespread mutagenicity in microbes suggests a strong *mutagenic potential.* This potential may not be expressed in other organisms because of efficient detoxification. The question remains whether all types of mammalian and all types of human cells detoxify efficiently enough or whether there are genetic variants which are less efficient? The problem of genetically determined differences in susceptibility to mutagenic/carcinogenic agents within the human race will be dealt with in the contribution of Unrau to this book (Chapter 12).

3.10 COMPLICATIONS IN TESTING COMPLEX MIXTURES

Testing complex environmental samples poses another problem. Such samples contain, in addition to known chemicals, many unidentified and even unknown chemicals. Some of these may in fact be mutagenic or carcinogenic and could be detected by themselves in a mutation test battery, but their activities may be masked by the presence of other chemicals which are acutely toxic to the test organisms. An environmental sample could contain enough of an acutely toxic chemical to kill all the indicator organisms before the active concentrations of the mutagenic/carcinogenic components are reached. Therefore, fractionation of samples before testing is critically important for a thorough analysis. A very elegant approach to solve this problem is described by Møller and Bjørseth in this book (Chapter 4).

3.11 DOSAGE SELECTION

Not only is it important to select the right test or tests, but the test conditions must be carefully selected. One major source of discrepancy is in the use of large and standard dose increments which do not pick up effects restricted to narrow dose ranges. This is especially important in the case of chemicals with a very sharp toxicity where a small increase in dose means a shift from no effects at all, not even toxicity, to killing all cells or individuals of the test organism. It is absolutely essential that dose increments be selected on the basis of the 'steepness' of the killing curve, not according to a fixed standard pattern. This can be illustrated by the induction of aneuploidy in yeast with γ-valerolactone (page 171). The lowest dose inducing aneuploidy was 2.20%; the peak of induction was found at 2.68%, the induction being greater than that found at 2.91%. A concentration of 4% killed all the cells under the chosen treatment conditions. Use of standard dose increments of 2% (not a wide range by most standards) would have resulted in failure to detect the relatively strong induction of aneuploidy which occurred between a concentration of 2%, which would have been too low for induction, and the completely lethal concentration of 4%.

3.12 EVALUATION OF DATA

Interpretation of the raw data may also be a source of discrepancies. What is a positive response? When is non-mutagenicity sufficiently well documented? Ideally, a test can give one of two results: positive or negative. However, four possibilities are present when it comes to predicting non-mutagenicity/non-carcinogenicity or mutagenicity/carcinogenicity. The positive response may be correct or false; the negative response may also be right or wrong. The terms 'correct positive', 'correct negative', 'false positive' and 'false negative' are only of operational value. A well documented 'false positive' is 3-methyl-4-nitro-quinoline-*N*-oxide, used as a mammalian non-carcinogen in the international trial mentioned above. 4-Nitroquinoline-*N*-oxide is an established carcinogen and one of the most potent mutagens, but its methyl derivative had been found negative in a carcinogenicity test where both chemicals were tested in the same dose range; the methyl derivative, however, was found to be active in most microbial mutation assays at doses a thousand times higher than those needed with 4-nitroquinoline-*N*-oxide [38]. It is easy to cover a wide dose range in a microbial short-term mutagenicity test, but it is very costly to do so in a test with whole mammals.

As already mentioned, false positive responses in microbial systems can be caused by an efficient detoxification system in mammals. False negatives will be created most frequently by inadequate metabolic activation. On the other hand, false negatives are possible when not all types of primary damage to DNA are processed into some detectable genetic effects. The cell type or organism used in a mutation test may not have an error-prone repair system acting on the specific kind of induced primary damage. Although undetected, the agent does indeed induce DNA damage and this type could lead to mutation in other organisms or cell types.

A classic false negative is an agent which induces only chromosomal mal-segregation but has been tested in assays based on the detection of genetic change initiated by primary damage inflicted on DNA, such as point mutation. A test battery should always include a test for aneuploidy, since the major target is not DNA but tubulin, one of the components of the spindle fibres which move chromosomes. No extensive screening of chemicals has been done for the induction of aneuploidy. Consequently, we still do not know how accurately the currently used mutation test batteries cover the spectrum of genetic events in humans. We do know that some mutagens have been missed by current tests because they induce aneuploidy but not other types of genetic effects, such as point mutation or chromosomal structural abberrations. (see Chapter 11). Evaluation of data obtained with test batteries not including assays for chromosomal mal-segregation will remain fraught with considerable uncertainty.

3.13 HOW WELL VALIDATED ARE MUTAGENICITY TESTS?

In the real testing condition we do not know whether to expect mutagenic activity or not. In academic conditions investigators often know what they have: a mutagen and a non-mutagen. It is common practice to modify experimental parameters until the known carcinogen behaves like a mutagen in the test used. Therefore, many tests which have been calibrated with a known set of mutagens/carcinogens have not been proven valid in predicting human mutagenicity–carcinogenicity. A test which repeatedly gives the right answer in blind testing is considered valid. A test can be calibrated with, for example, 50 mutagens and 50 non-mutagens. It may detect three non-mutagens as mutagens, which is a low error rate. Since, however, there is a tendency in calibration of a test to use many more mutagen/carcinogens than non-mutagen/non-carcinogens, another problem arises. If 950 non-mutagens and only 50 mutagens are to be tested, a realistic situation, an error rate of 6% would give by extrapolation 57 false positives. If we further assume that the test correctly identifies all 50 established mutagens, a less realistic result, we must conclude that more than half of the chemicals identified as mutagens are false positives. This makes the test seem invalid, but there is probably no test yet which has been validated so thoroughly. In other words, we really do not known how good tests are. Whatever they tell us, we must treat the results with caution and discretion.

3.14 CONTROLS, REPRODUCIBILITY AND DOSE RESPONSE CURVES

When we are dealing with the interpretation of positive or negative responses, a very important prerequisite is to have reproducible control values to be used as reference points. Reproducibility in controls allows detection of low levels of mutation induction and provides proof that the test organisms were responding properly. Such control values also have to be within the normal range of the control values obtained by the laboratory, and also, if available, those from other laboratories. Zimmermann and Scheel [39] reported on a case where the control values for mitotic gene conversion in yeast were lower than the typical level established in previous experiments, because the culture used consisted mostly of homoallelic cells in which mitotic gene conversion could no longer be detected.

The opposite problem of exceedingly high control values is common in tests which are based on the mutational or recombinational restoration of a function. A histidine-requiring microbial mutant can only grow when histidine is supplied in the medium. A non-requiring revertant will have a growth advantage under certain conditions, especially if the cultures are grown into stationary phase. The result will be that the revertant becomes more frequent not because of induction of mutation but because of selection. Let us assume that the control value for histidine reverse mutants at the end of a 24-hr growth period is 8.7×10^{-7} in one experiment and 38.6×10^{-7} in another. Let us further assume that exposure of the two cultures to a mutagenic agent leads to the induction of reverse mutants at a frequency of 17.4×10^{-7}. We can now add the induced frequency to the two different controls: in the first case where we have a control of 8.7×10^{-7} the total frequency combining both types of revertants – those of spontaneous and of induced origin – would be 26.1×10^{-7}. This is almost three times that for the control culture and could be considered as a indication of mutation induction. However, if the induced frequency of 17.4×10^{-7} is added to a control value of 38.6×10^{-7} the sum will be only 56.0×10^{-7}, and the increase over the control only 45%, a result which would not be taken as indicative of mutation induction. This example is taken from a testing report.

Erratic control values invalidate experiments. They signal that some essential parameter is not 'under control'. Most mutagenicity tests are done with growing cells because these are much more sensitive than non-dividing ones. If the growth conditions for controls and test samples are not identical, differences between them may not be due to the presence or absence of a potentially mutagenic agent. Erratic control values are usually associated with erratic positive responses. Simple statistics may even attribute significance to such differences between control and experimental values. However, the difference, no matter how significant, cannot be attributed to induction of mutation and could be caused by unknown and uncontrolled factors favouring selection of prototrophs. Statistics allows establishment of the significance of differences, but does not necessarily identify the causes of the differences.

Good evidence for induction of mutation is a smooth dose-response curve. This can be a straight line, or exponential, plateau or even bell-shaped. Still more complex curves have been reported, in which a peak value was followed by a second rise. The latter category might be found in growing cultures where some cells are in a highly mutagen-sensitive phase, such as DNA replication. Although they mutate easily they are also killed off much faster than cells in a more resistant stage of the cell cycle, which may be not only more resistant to killing but also more refractory to mutation induction. The population composition will change with increasing dose so that almost all the remaining cells are less mutable. This will result in a dose-response curve for mutation induction which rises sharply at low doses but drops at higher doses as the more sensitive cells are killed. As doses increase still further, mutation induction will become detectable in the surviving fraction, resulting in a second rise [48]. Such a situation requires many more dose levels than a simple curve does, in order to show that the response, no matter how complex, is continuous and not merely erratic.

3.15 CONCLUSION

Having considered many of the problems involved in selecting a mutagenicity test battery, the investigator is still faced with making a selection. At present the recommendation would be to select a test which has been used in many laboratories to test many chemicals and complex mixtures. No matter how good a test is, either on theoretical grounds, in human tissue-culture cells, or in *in vivo* tests in mammals closely related to man in behaviour, it has to stand the test of actual mutagenicity screening.

It is absolutely mandatory to understand the mutagenicity test which is used. A test performed by the expert looks easy and simple. But all biological tests require some experience. Microbial and tissue-culture cell tests require familiarity with sterile working procedures and possible sources of contamination. Cytogenetic assays can only be used by persons with experience in microscopy. Whole animal mammalian tests are quite demanding and responses may change depending on conditions of feeding and infectious diseases. The complexity of even a unicellular organism provides many sources of potential error. It is advisable to become acquainted with a given test in a laboratory where it has been used successfully and has generated reliable data.

It must be mentioned finally that the choice of tests and their use have not always been motivated by intensive consideration of their genetic-biological merits. Rather, tests have often become available as the results of other activities not directly related to mutagenicity testing, and have become popular owing to promotional skill. There is still ample room for new developments.

Of course, there is no excuse for waiting for the ultimate, validated test system. The problems facing us are urgent, and we have to take what is available and, providing that the data obtained are judiciuosly interpreted, we can learn much and take appropriate measures to reduce environmental mutagenic pollution.

GENERAL REFERENCES

A very useful presentation of various aspects of mutation and mutagenesis in respect to humans can be found in the text book by E. Novitski, *Human Genetics*, 2nd Ed., Macmillan, New York, 1982.

The journal *Mutation Research*, Elsevier, Amsterdam, contains a wide collection of articles dealing with mutagenesis and related subjects, covering the full range of mutational effects in a wide range of organisms.

REFERENCES

[1] H. J. Muller, *Z. Ind. Abst. Vererbungsl.*, 1927, **1** (Suppl.), 234.
[2] C. Auerbach, *Mutation Research: Problems, Results and Perspectives*, Chapman & Hall, London, 1976.
[3] H. H. Smith and A. M. Srb, *Science*, 1951, **114**, 490.
[4] A. M. Clark, *Z. Vererbungsl.*, 1960, **91**, 74.
[5] *IARC Monographs on the Evaluation of Carcinogenic Risk of Chemicals to Man*, Vol. 10, Lyon, 1976.
[6] K. H. Bauer, *Das Krebsproblem*, 2nd Ed., Springer, Berlin, 1963.
[7] H. Marquardt, F. K. Zimmerman, H. Dannenberg, H.-G. Neumann, A. Bodenberger and H. Metzler, *Z. Krebsforsch.*, 1970, **74**, 412.

[8] F. K. Zimmerman, *Biochem. Pharmacol.*, 1971, **20**, 985.
[9] B. N. Ames, W. E. Durston, E. Yamasaki and F. D. Lee, *Proc. Natl. Acad. Sci. U.S.A.*, 1973, **70**, 2281.
[10] B. N. Ames, *Science*, 1983, **221**, 1256.
[11] J. R. S. Fincham, *Genetics*, Wright, Bristol, 1983.
[12] M. Pirastu, L. del Senno, F. Conconi, C. Vullo and Y. W. Kan, *Nature*, 1984, **307**, 76.
[13] R. A. Spritz, P. Jagadeeswaran, P. V. Choudary, P. A. Biro, J. T. Elder, J. K. de Riel,, J. L. Manley, M. L. Gelter, B. Forget and S. A. Weissmann, *Proc. Natl. Acad. Sci. U.S.A.*, 1981, **78**, 2455.
[14] E. B. Hook, *Mutation Res.*, 1983, **114**, 389.
[15] H. Fraenkel-Conrat and P. C. Kimball, *Virology*, Prentice Hall, Englewood Cliffs, 1982.
[16] B. Heidemann, in *Chemical Mutagens: Principles and Methods for their Detection*, Vol. 1, A. Hollaender (ed.), p. 235, Plenum Press, New York, 1971.
[17] C. Stern, *Genetics*, 1936, **21**, 625.
[18] L. B. Russell and M. H. Major, *Genetics*, 1957, **42**, 161.
[19] R. Fahrig, *Molec. Gen. Genet.*, 1984, **194**, 7.
[20] E. Käfer, P. Marshall and G. Cohen, *Mutation Res.*, 1976, **38**, 141.
[21] F. K. Zimmerman, in *Handbook of Mutagenicity*, 2nd Ed., B. J. Kilbey, M. Legator, W. Nicholas and C. Ramel (eds.), Elsevier, Amsterdam, 1984, p. 215.
[22] B. K. Vig, *Mutation Res.*, 1975, **31**, 49.
[23] M. Radman, C. Dohet, M. F. Bourgignon, O. P. Doubleday and P. Lecompte, in *Chromosome Damage and Repair*, E. Seeberg and K. Kleppe (eds.), Plenum Press, New York, 1981, pp. 431–445.
[24] T. Lindahl, *Ann. Rev. Biochem.*, 1982, **51**, 61.
[25] B. K. Duncan and J. H. Miller, *Nature*, 1980, **287**, 560.
[26] L. Prakash, *Genetics*, 1974, **78**, 1101.
[27] R. H. Haynes and B. A. Kunz, in *The Molecular Biology of the Yeast* Saccharomyces cerevisiae: *Life Cycle and Inheritance*, J. N. Strathern, E. W. Jones and J. R. Broach (eds.), Cold Spring Harbor Laboratories, 1981, p. 371.
[28] J. C. Game, in *Yeast Genetics – Fundamental and Applied Aspects*, J. F. T. Spencer, D. M. Spencer and A. R. W. Smith (eds.), Springer, New York, 1983, p. 109.
[29] R. Holliday and J. E. Pugh, *Science*, 1975, **187**, 226.
[30] R. L. P. Adams, T. Davis, J. Fulton, D. Kirk, M. Qureshi and R. H. Burdon, *Current Top. Microbiol. Immunol.*, 1984, **108**, 143.
[31] W. Doerfler, *Ann. Rev. Biochem.*, 1983, **52**, 93.
[32] W. Doerfler, *Current Top. Microbiol. Immunol.*, 1984, **108**, 79.
[33] J. K. Christman, P. Price, L. Pedrinan and G. Acs, *Eur. J. Biochem.*, 1977, **81**, 53.
[34] S. M. Taylor, P. A. Constantinides and P. A. Jones, *Curr. Top. Microbiol. Immunol.*, 1984, **108**, 115.
[35] P. G. Constantinides, P. A. Jones and W. Gevers, *Nature*, 1977, **267**, 364.
[36] P. A. Jones, S. M. Taylor, T. Mohandas and L. J. Shapiro, *Proc. Natl. Acad. Sci. U.S.A.*, 1982, **79**, 1215.
[37] M. Harris, *Cell*, 1982, **29**, 483.
[38] F. J. de Serres, and J. Ashby (eds.), *Evaluation of Short-term Tests for Carcinogens*, Elsevier/North Holland, New York, 1981.
[39] F. K. Zimmerman and I. Scheel, *Mutation Res.*, 1984, **139**, 21.
[40] V. L. Wilson, and P. A. Jones, *Cell*, 1983, **32**, 239.
[41] R. Holliday, *Br. J. Cancer*, 1979, **40**, 513.
[42] I. Purchase, *Mutation Res.*, 1982, **99**, 53.
[43] B. J. Kilbey, M. Legator, W. Nicholas and C. Ramel (eds.), *Handbook of Mutagenicity Test Procedures*, 2nd Ed., Elsevier, Amsterdam, 1984.
[44] E. Vogel and A. T. Natarajan, *Mutation Res.*, 1979, **62**, 51.
[45] H. V. Malling, *Mutation Res.*, 1971, **13**, 537.
[46] G. R. Mohn and J. Ellenberger, *Mutation Res.*, 1973, **19**, 257.
[47] C. Ramel, J. Drake and T. Sugimura, *Mutation Res.*, 1980, **76**, 297.
[48] F. K. Zimmerman and R. Schwaier, *Z. Vererbungsl.*, 1963, **94**, 261.

4

Chemical separation and *in situ* mutagenicity testing

M. MØLLER and **A. BJØRSETH,** Central Institute for Industrial Research, P.O. Box 350, Blindern, Oslo 3, Norway
VIRGINIA S. HOUK, U.S. Environmental Protection Agency Research Center, Research Triangle Park, NC 27711, U.S.A.

4.1 INTRODUCTION

Studies over the last decade have clearly demonstrated that compounds causing genotoxic effects are widespread in the environment. Much of the recent effort in environmental chemistry and toxicology has been directed towards characterizing these genotoxic compounds (for a review, see references [1] and [2]). The traditional way to characterize a complex sample has been by chemical analysis to identify the individual components. However, by this approach unknown or overlooked mutagens or carcinogens may remain undetected. In recent years a combination of chemical analysis and mutagenicity testing has been applied successfully to fractionated samples for identification of mutagens in complex mixtures. By repetition of chemical fractionation and biological testing, subfractions with relatively high concentrations of mutagenic compounds may be worked up for the final determination of the constituents by chemical analysis. However, such methods may involve considerable work and expense.

In a recent paper [3] we reported a new technique in which complex samples were separated on thin-layer chromatography plates and their mutagenic effect was registered directly on the plates by the *Salmonella* mutagenicity assay. This method is in principle an extension and further improvement of a technique previously reported by Commoner [4]. A sample of 2-acetylaminofluorene (AAF) was applied to a TLC-strip developed conventionally and the strip was subjected to an Ames test. A cluster of colonies appeared at the location of the AAF zone, demonstrating that the mutagen (AAF) could be released from the TLC plate and that the *Salmonella* bacteria may act as detectors of mutagens on the plate [4].

The TLC/*Salmonella* technique has several advantages. Complex samples can be separated by one-dimensional or two-dimensional thin-layer chromatography before exposure to the bacteria. The method is quick and simple to use and is particularly useful for screening large numbers of samples. Many complex samples contain toxic compounds which may interfere with the mutagenicity assay, giving misleading results, but by use of the TLC plates, toxic compounds or components with synergist/antagonistic effects may be separated from the mutagens and detected independently. Another advantage of this method is the ease of qualitative comparison of mutagens in different samples. For identification of the mutagens detected, extracts from duplicate plates can be analysed chemically.

This chapter gives an outline of the TLC/*Salmonella* technique and includes a discussion of its advantages and disadvantages. The application of the method to different complex samples is included, to illustrate the usefulness of the technique.

4.2 EXPERIMENTAL METHODS

4.2.1 Thin-layer chromatography (TLC)

TLC is a well-established separation method for characterization of complex samples. Its advantages are the mild treatment of sensitive compounds, the simplicity of the equipment needed and the possibilities for variations of plate material and solvent systems.

When different types of components are to be separated, the chromatography can be performed on plates coated with different adsorbent materials. Non-polar lipophilic compounds should be separated on plates coated with silica or alumina, while cellulose-coated plates give a better separation for moderately polar compounds. Hydrophilic, water-soluble components are best fractionated on cellulose plates or by reversed-phase chromatography.

4.2.2 Pretreatment of TLC plates

The TLC plates used were obtained commercially and had silica, cellulose and reversed-phase coatings. The 10 × 10 cm plates used in our studies were developed in absolute ethanol and dried for 5 min at room temperature and then for 20 min at 80°C. The plates were then assumed to be sterile.

4.2.3 Development of plates

The samples (1–5 μl) in acetone were spotted onto the plates and the separation was done with various solvent systems at room temperature in a 12 × 12 cm chamber. Separated components were made visible by viewing under ultraviolet light (254 and 350 nm) or on duplicate plates by conventional (spraying) techniques [5].

4.2.4 TLC/Salmonella mutagenicity assay

The *Salmonella typhimurium* strains TA 98 and TA 100 (containing the R-plasmid) have been shown to be the most sensitive strains for the testing of many environmental samples and these two strains were chosen for studies using the TLC/

Salmonella-assay. The strains were kindly supplied by B. N. Ames, Berkely, Ca. Only the expected low number of spontaneous revertants are observed on blank TLC plates.

The mutagenicity assay was performed essentially as described by Ames *et al.* [6], by using TA 98 and TA 100 with or without S9 mix prepared from rat liver (pretreated with Aroclor 1254).

The developed TLC plates were placed in the bottom of Petri dishes (14.5-cm diameter), and minimal agar (30 ml) was carefully added to each dish. During this and the subsequent operations care was taken to avoid damaging the surface of the TLC plate. The liquid agar was poured into a part of the bottom of the dish not covered by the TLC plate. The agar was then slowly distributed over the surface of the plate. It was essential that the minimal agar layer should be as thin as possible in order to get good contact between the components on the plates and the bacteria.

After 1-2 hr the tester strain (0.3 ml) with or without S9 mix (1.5 ml) was mixed with 6 ml of top agar, and poured on to the minimal agar. After 2-3 days at 37°C, clusters of histidine revertants were observed in the areas containing mutagens (3 days incubation results in larger colonies that can more easily be registered).

Variations in the layers in contact with the TLC plate

In order to get the most sensitive system for detecting mutagens on the TLC plates, different combinations of minimal agar (MA), top agar (TA) and TLC plate have been tried. Solutions of mutagenic compounds were spotted onto the TLC plates, the solvent was evaporated and the plates were subsequently tested with various agar combinations. The following combinations were used (from top to bottom): TA, MA, TLC (normal); MA, TA, TLC; (MA + TA mixed), TLC; and TLC, TA, MA. The results showed that the system (MA + TA mix), TLC, was as sensitive as the system adopted, of TA, MA, TLC. However, we recommend using separate top agar and minimal agar layers because the colony scoring in this system is often easier to do because the colonies are bigger. Another advantage is that the compounds can diffuse into the agar before the bacteria are added in the top agar. The combinations TLC, TA, MA and MA, TA, TLC were found to be less successful.

Diffusion into the agar layer

To generate a mutagenic effect, the compound must be released from the TLC adsorbent and diffuse into the agar. The kinetics of this process have not been very much studied, but sufficient time must be allowed for the compounds to reach the bacterial cells while the growth factors and S9 enzymes for activation are still available. Special attention should be given to diffusion of mutagens into the agar. This process can be studied by using coloured (non-mutagenic) compounds with chemical and physical properties similar to the compounds in question. These may be spotted onto the plate and subsequently treated as above. The colour readily shows how these compounds are mobilized and diffuse into the agar layer. This has been done with several commercially available compounds, including Merck Lipophilic Test Mixture (Indophenol Blue, Sudan Red G, 4-dimethylaminoazobenzene). Compounds of different polarity were chosen for testing the system with mutagens. The compounds were spotted onto the TLC plates without chromatography. Table 4.1 shows the mutagens tested and the results obtained. The direct alkylating agent sodium azide, which is very polar, gave a positive result in TA 100, but it diffused over a large area and did

Table 4.1 – Confirmed mutagens tested by the TLC/Ames test. From [22], reprinted by permission of the authors and copyright holders.

Sodium azide	Streptozotocin
Methyl methanesulphonate	4-Nitroquinoline-*N*-oxide
3-Propiolactone	MNNG
2-Nitrofluorene	2-Anthramine
Cyclophosphamide	Benzo(*a*)pyrene
2-Acetylaminofluorene	Epichlorohydrin
Daunomycin	Dimethylcarbamyl chloride

not give colonies in one distinct area. Our experience is that compounds of intermediate polarity (such as 2-aminoanthracene and diaminoanisole) are the most suitable for detection in the TLC/*Salmonella* assay, giving nice clusters of colonies around the spot. The non-polar compounds benzo(*a*)pyrene and benz(*a*)anthracene gave negative results in this spot test. In order to overcome this problem several modifications of the test method have been tried (see below).

Testing of non-polar compounds

It has been pointed out by Ames *et al.* [6, 7] that most polycylic hydrocarbons (PAH) do not give a mutagenic response in a spot test, i.e. the compounds do not diffuse into the agar, and hence do not reach the bacteria.

In our study no mutagenic effect was observed when various non-polar compounds, including benzo(*a*)pyrene (BaP) and benz(*a*)anthracene were spotted onto TLC plates and the agar and the *Salmonella* bacteria were added directly on top. Furthermore, there was no difference between cellulose and silica plates. In order to get the non-polar compounds into the agar, the technique was modified by addition of surfactants or solvents. Inclusion of 1% of Tween 80 or a commercial washing detergent in the agar did not influence the solubility of BaP. Neither did up to 20% of DMSO or 0.2% of lecithin in the agar result in any mutagenic response.

The problem of failure of the mutagenicity test with non-polar compounds has been experienced by others [8]. On the basis of these experiments and our experience with environmental samples, we conclude that the TLC/*Salmonella* assay at the present stage of development is primarily suitable for compounds of intermediate polarity.

4.3 APPLICATIONS

In principle, *in situ* mutagenicity testing may be applied to a large variety of samples from different environmental matrices. The method has been useful for detection of mutagens in extracts of ambient air [3], extracts of typewriter ribbons [9], emission samples [10] and industrial discharges [11]. However, we will limit our discussion here to some applications in the areas of aqueous discharges, product control, emission samples and hazardous waste.

4.3.1 Aqueous discharges

Aqueous discharges from many types of industry are found to contain mutagens which may give rise to environmental pollution problems. Effluent samples from two such industries have been tested by the TLC/*Salmonella* assay in order to characterize the mutagenic impurities.

Weaving industry effluents

A Swedish research project screened several industrial waste water samples for their biological activity by using a number of methods, including the Ames mutagenicity test [12]. One sample of particular interest was waste water from a textile plant. This waste water contains a large number of chemicals which are used as carriers and additives aimed at improving products and processes.

An extract of the effluent from a textile plant that was tested by the Ames test showed a strong direct mutagenic effect on strain TA 98 [11]. The activity decreased by approximately 50% on addition of S9 mix. This extract, an acid/ neutral fraction of a total extract, was separated on TLC plates and tested *in situ* for mutagenic properties. It was shown that the extract contained three mutagenic areas with clusters of revertant colonies, as shown in Fig. 4.1.

A mutagenic area near the the application point (i.e. containing polar compounds) showed the highest mutagenic activity in the presence of S9 while the area having an R_F value of about 0.2 in elution system I and 0.3 in system II showed the highest activity in the absence of S9. This clearly demonstrates that, by this technique, it is possible to detect simultaneously compounds that have different activation properties.

Pulp-bleaching effluents

It is well known that effluents from the pulp and paper industry may cause toxic effects in aquatic organisms. Of particular environmental interest are effluents from chlorine bleacheries. During the bleaching sequences, large amounts of chlorinated organic compounds are formed [13]. The environmental impact of chlorination has been demonstrated by Pfister and Sjöström [14] who showed that effluents from the chlorination stage are up to 10 times more toxic to fish than the effluents from other stages of bleaching.

Effluents from chlorine bleacheries also contain compounds exhibiting mutagenic and potential carcinogenic effects [15], including direct-acting mutagens [16]. For these kinds of sample TA 100 has been shown to be the most sensitive strain for detection of mutagens [13]. Many of the identified mutagenic compounds are chlorinated compounds of low molecular weight,

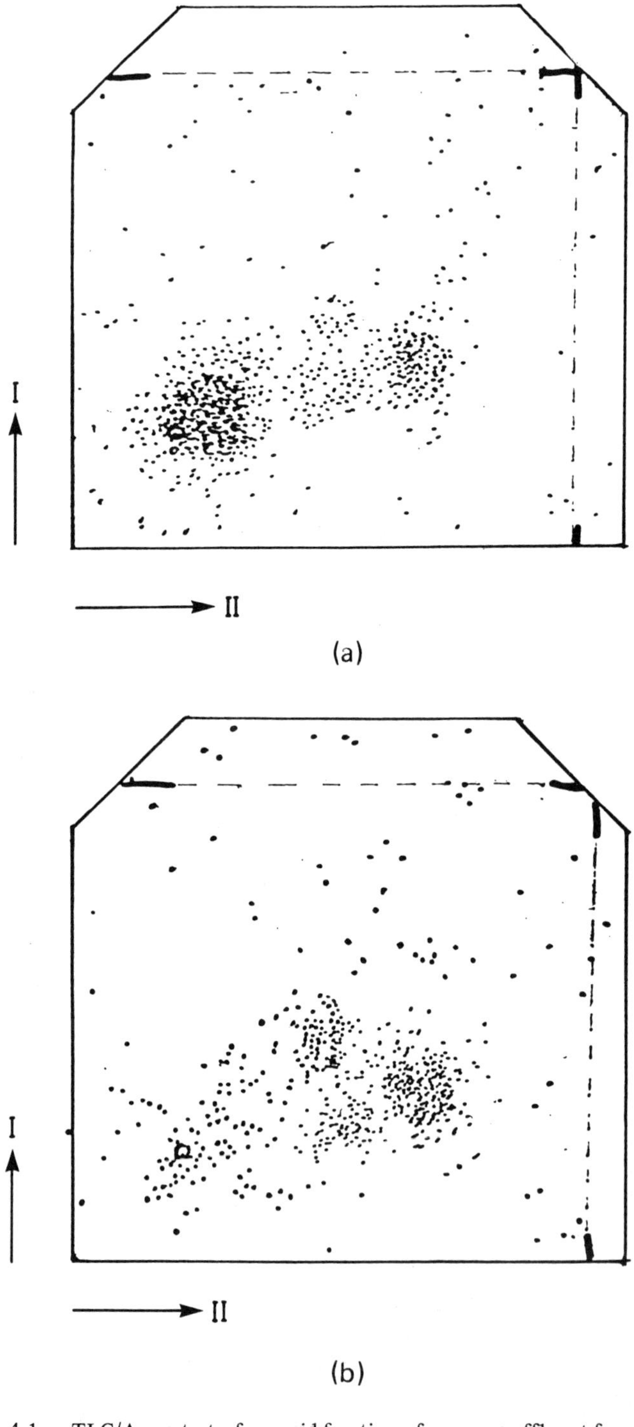

Fig. 4.1 – TLC/Ames test of an acid fraction of aqueous effluent from a weaving mill, with use of strain TA 98. (a) is with S9 and (b) is without S9.

having relatively polar properties [17]. In order to study the content of mutagens in effluents from different bleaching processes in more detail, extracts of bleaching effluents from a sulphite- and a kraft-pulp plant were first spotted onto a TLC plate and then subjected to a *Salmonella* test. The results are shown in Fig. 4.2. As can be seen, only the extracts from the kraft pulp showed a clear mutagenic response. However, the results also revealed that this extract contained toxic compounds which left a zone with no bacterial growth around the application point.

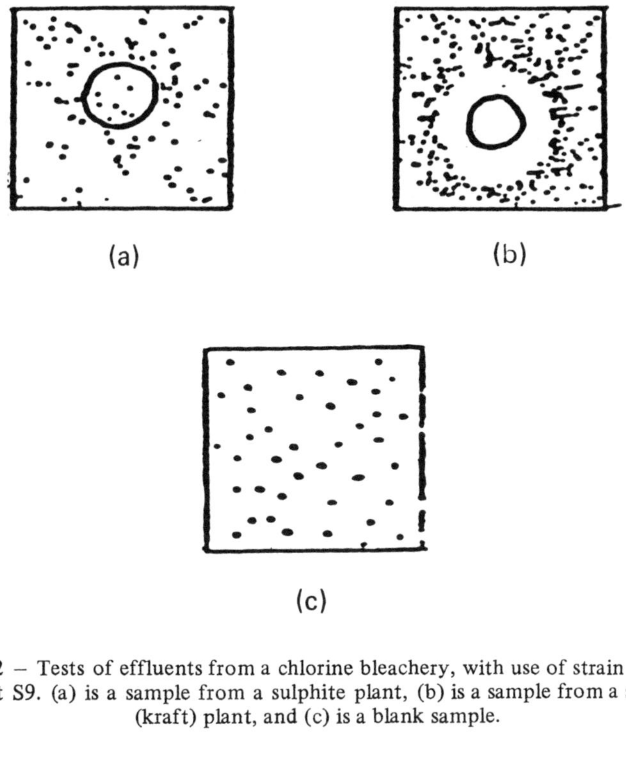

Fig. 4.2 – Tests of effluents from a chlorine bleachery, with use of strain TA 100 without S9. (a) is a sample from a sulphite plant, (b) is a sample from a sulphate (kraft) plant, and (c) is a blank sample.

Extracts from the chlorination stage of the kraft pulp were further chromatographed on TLC plates and tested for appearance of mutagenic areas. The results showed that, with the usual elution system, both mutagens and toxic compounds remained near the application point (Fig. 4.3 a and b). By modification of the elution systems, improved separation of the compounds in both elution directions was achieved. However, the toxic area increased considerably in size and no distinct areas with mutagens were obtained, except near the application point (Fig. 4.3 c and d). These experiments indicate that the TLC/ *Salmonella* system has limited applicability for pulp-bleaching samples.

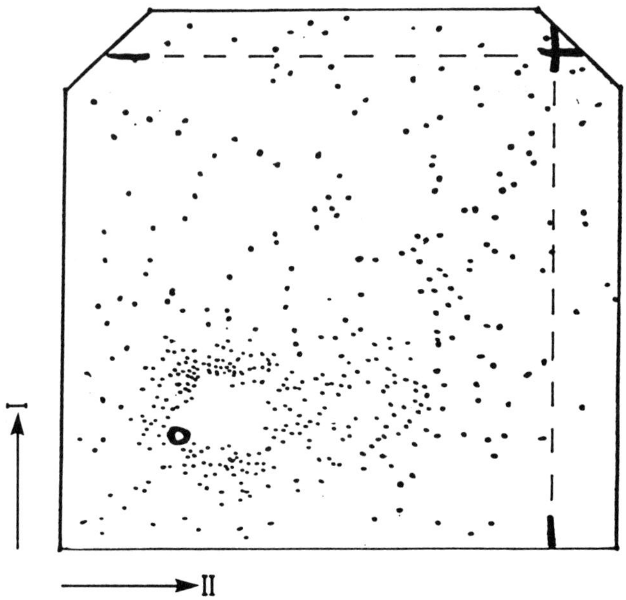

(a)

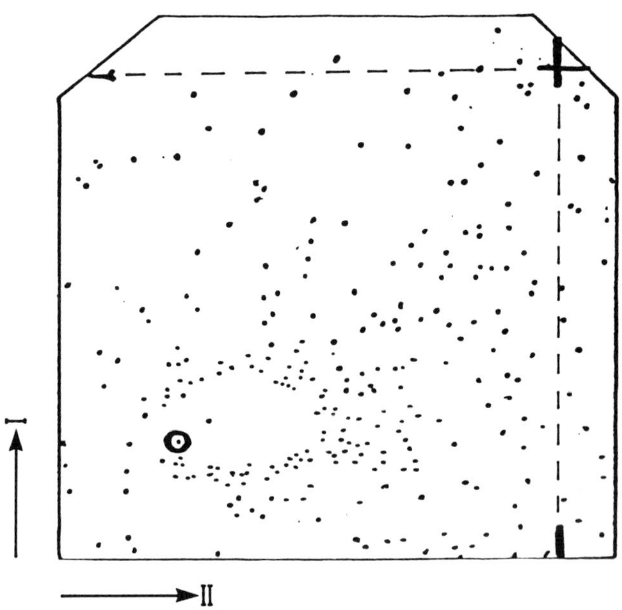

(b)

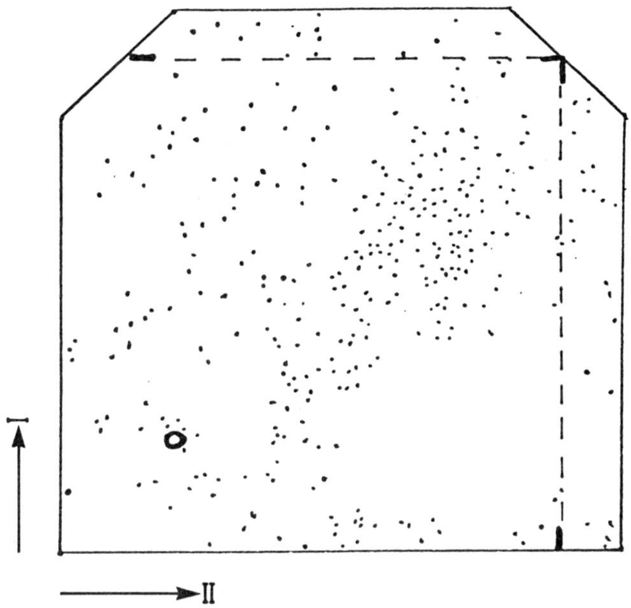

(c)

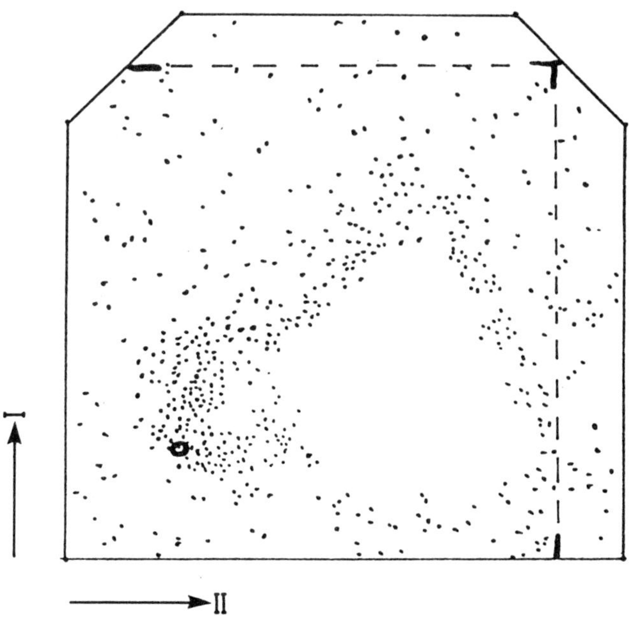

(d)

Fig. 4.3 – TLC/Ames test of effluents from a sulphate (kraft) plant with use of TA 100 without S9. (a) is a sample from the sequence alkali/ozone/water/chlorine while (b) is a sample from alkali/oxygen/chlorine, both using a standard elution system [9]. Samples (c) and (d) correspond to (a) and (b) respectively, but a modified elution system has been utilized (chloroform/ethyl acetate/acetone/ acetic acid, 60:30:8:2 v/v) [17]. (○–Point of application).

4.3.2 Product control

A very important task in consumer product control (as well as in many scientific studies) is to investigate the purity of commercial products. This is important both for consumer health and safety and for validity of the scientific investigations.

One of the advantages of the TLC/*Salmonella* assay is the easy detection of mutagenic impurities in a product such as a chemical compound. Figure 4.4 demonstrates how the compound 1-aminopyrene (NH_2-P) gives rise to two mutagenic areas, both in the presence and absence of S9. Area I shows the mutagenic effect of NH_2-P itself. It can be seen from the lack of growth in the centre of this area that, in high doses, NH_2-P is toxic to the bacteria. The impurity in the product gives rise to mutagenic area II. On the basis of its chromotography properties, it is predicted that this mutagenic impurity is a less polar compound. Addition of S9 increased the mutagenicity.

We have earlier demonstrated by this technique that a specimen of the compound 2,4-diaminoanisole contained several impurities [3]. As mentioned above, typewriter ribbons and carbon paper have been shown to contain 2 or 3 classes of mutagenic impurities [9]. A large number of extracts from typewriter ribbons were studied to show the usefulness of the method for qualitative comparisons of mutagens in the samples [9]. Three main areas containing mutagens could be separated on the TLC plates used for these extracts. Some extracts gave one mutagenic area while others gave two or three. The mutagenic response in the presence or absence of S9 also varied from extract to extract and could be clearly seen on the TLC plates.

4.3.3 Emission samples

Various types of emission samples, such as fly-ash from coal combustion and diesel exhaust have been shown to contain mutagens [18, 19]. Other industrial emissions contain mutagenic material, particularly where the process involves high temperature. For mutagenicity testing and analysis purposes these samples are often collected as three fractions: the particulates (collected on filters), a condensate, and finally the more volatile components collected on an XAD-2 column.

It has been demonstrated that emission samples may contain toxic compounds which can interfere with mutagenicity testing [10], causing misleading results, in particular if the toxic compounds have chemical properties similar to those of the mutagens in the sample. Testing the mixture without any separation may give results in which the mutagenic effect is not detected or is significantly reduced compared to the actual biological activity.

A marked advantage of the TLC/*Salmonella* technique is that the compounds in a mixture are separated on the TLC plate before the bioassay, and may be especially useful for testing emission samples.

Emissions from graphite production

Production of graphite is an industrial process involving high temperatures and organic material. The results from testing emissions from such a plant are given in Fig. 4.5.

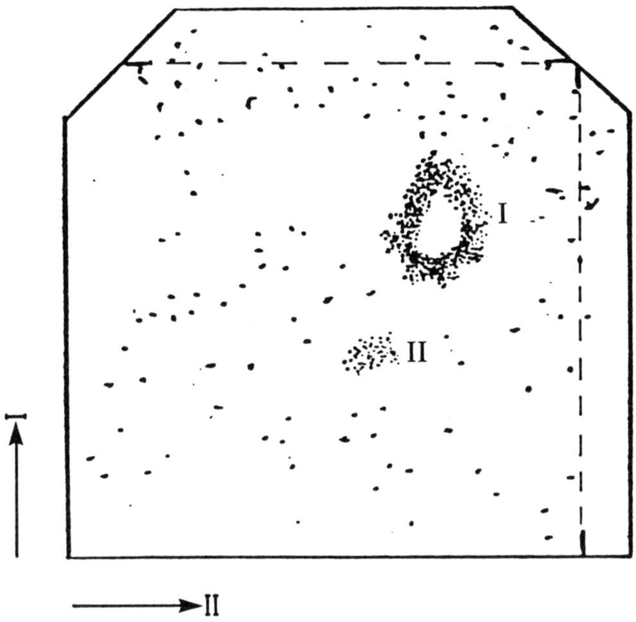

(a)

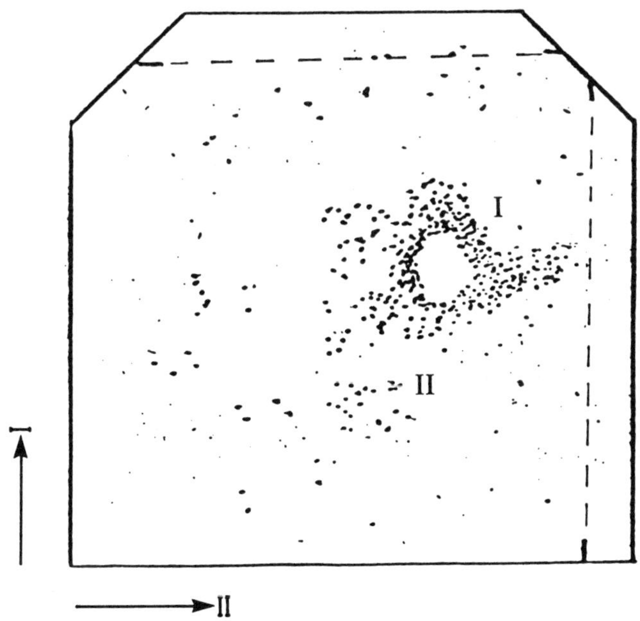

(b)

Fig. 4.4 – TLC/Ames test of commercial 1-aminopyrene with use of strain TA 98 (a) with and (b) without S9 activation. Solvent systems as modified in [11].

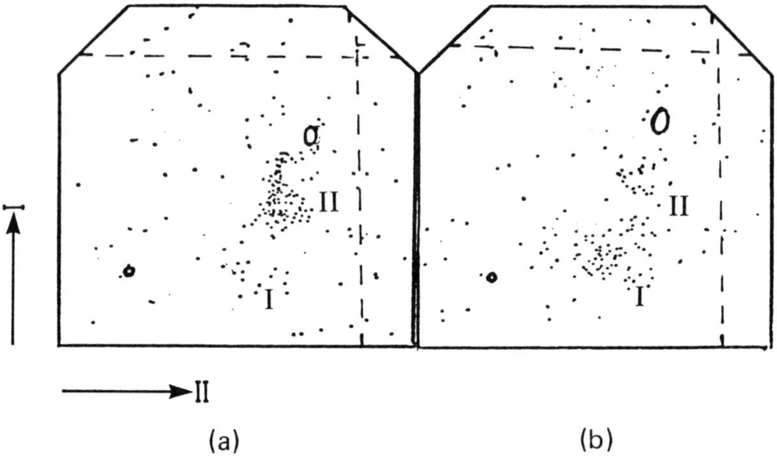

Fig. 4.5 – TLC/Ames test of emissions from a graphite plant using TA 98 and S9 activation. (a) shows extracts of particulate matter while (b) shows an extract of a condensate and XAD-2 trap. Solvent systems as in [9].

Fig. 4.5 (a) shows the mutagenic pattern of the particulate fraction in the presence of S9. Two mutagenic areas appear on the plate. The PAH compounds can be seen on the plate as a bright yellow spot but there is no indication of mutagenic activity in this area. Figure 4.5 (b) shows the mutagenic activity of the more volatile fraction in the presence of S9. This fraction contains only one mutagenic area (I) in addition to the non-mutagenic PAH area. Thus in addition to PAH there must be at least one or two classes of mutagens that contribute to the mutagenic activity of the extracts.

Emissions from burning of cereal straw

Because of rising oil prices, there has been an increase in the use of wood, straw etc., as fuels. Recent studies have shown high concentrations of PAHs and high mutagenic activity in emissions from combustion of wood and cereal straw [20, 21].

The mutagens in cereal straw samples were studied in detail, by use of the TLC/*Salmonella* system. The results are presented in Fig. 4.6. In the presence of S9 the particulate fraction showed mutagenic activity in three areas, I, II, III (Fig. 4.6 a). However, in the absence of S9 (Fig. 4.6 b) this extract was clearly toxic and even the background lawn of bacteria disappeared in area II.

No mutagenic activity was observed in the XAD-2 extracts with or without S9 (Fig. 4.6 e and f). So far the compounds responsible for the mutagenic effects have not been identified, but the mutagens in area I may possibly be substituted PAHs.

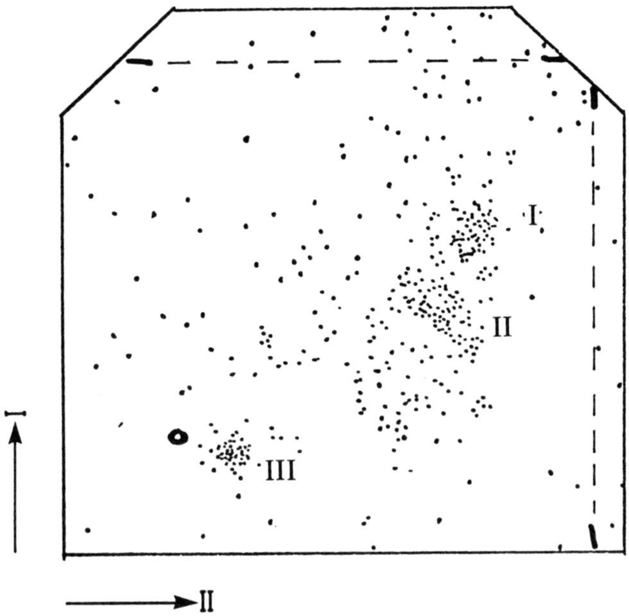

(a)

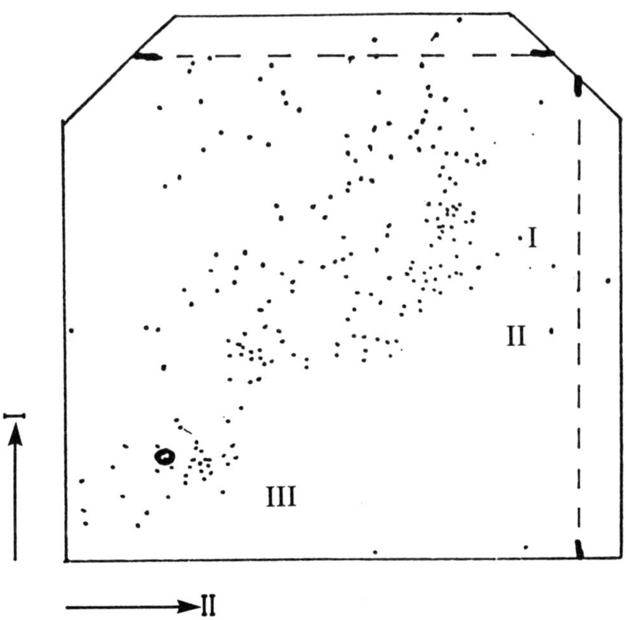

(b)

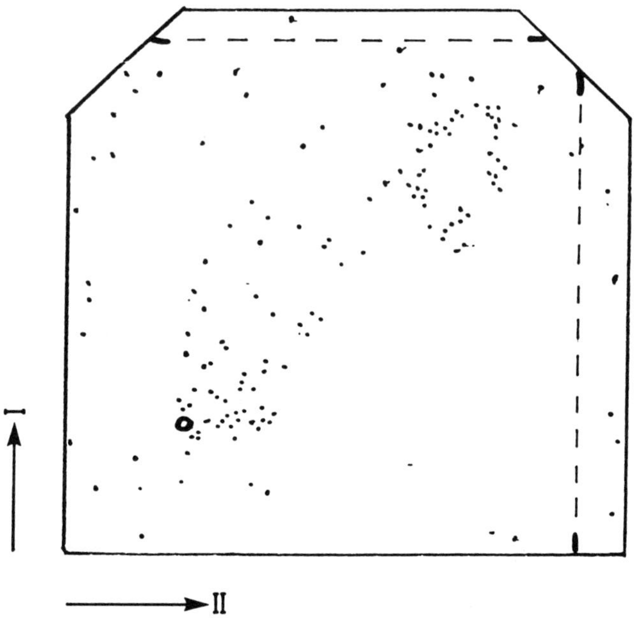

(c)

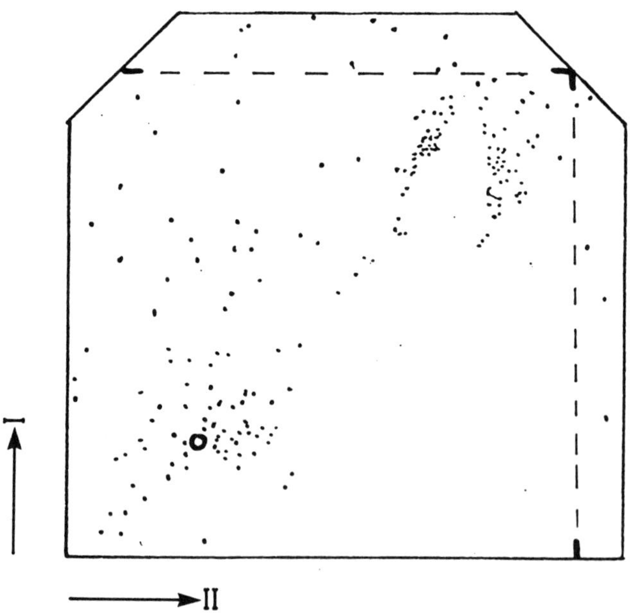

(d)

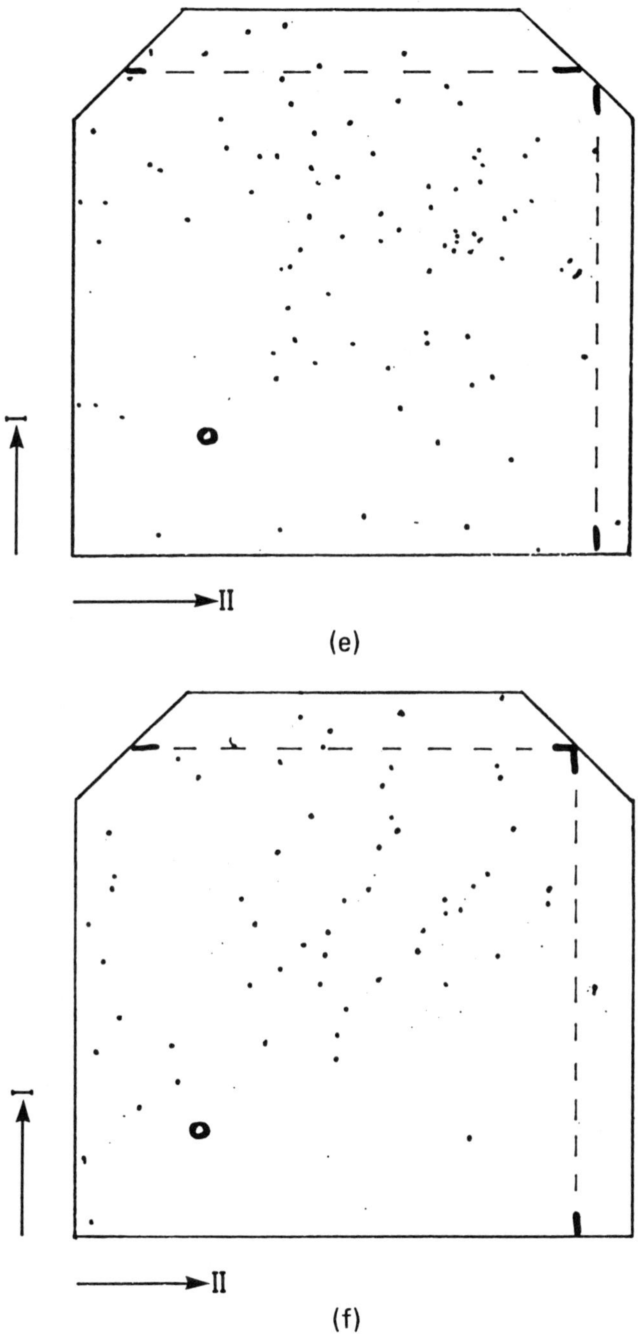

Fig. 4.6 – TLC/Ames test of emissions from combustion of cereal straw, with use of strain TA 98. The samples are extracts of particles (a) with and (b) without S9, condensates (c) with and (d) without S9, and XAD-2 traps (e) with and (f) without S9. Solvent systems as modified in [11].

4.3.4 Hazardous waste

Another area of considerable environmental importance is hazardous waste. The TLC/*Salmonella* assay has recently been applied to characterize a number of hazardous waste sites in the U.S. [8, 22] and some results will be summarized here. In addition, some recognized mutagens have been tested by the assay to confirm its feasibility.

Table 4.1 shows the mutagens tested. All except benzo(*a*)pyrene and two chlorinated organic compounds gave a positive response. The results for benzo(*a*)-pyrene completely confirm results reported previously [3]. In addition, one non-mutagen (1-napthylamine) was tested and gave negative results.

Table 4.2 shows some results of TLC/*Salmonella* assays of hazardous waste samples [8, 23], and Table 4.3 provides additional information about the samples. All samples described in Table 4.2 were tested as received, except for samples 5, 6, 12, 13, 14 and 16, for which extracts were tested. Table 4.2 illustrates three kinds of response which were accepted as positive.

(1) A reproducible two-fold increase in spontaneous revertant (negative control) counts;
(2) a non-random clustering of colonies occurring at localized spots on the plate;
(3) a random cluster effect which could be detected only with the aid of a grid and numerical comparison of revertant numbers per unit area.

Table 4.2 – Testing of hazardous wastes. (From [22], reprinted by permission of the authors and copyright holders).

WASTE SAMPLE	TLC PLATE DOSE RANGE μl		TLC/AMES ACTIVITY TA 98				TA 100			
	TA 98	TA 100	+S9	R_F	−S9	R_F	+S9	R_F	−S9	R_F
1. Organic Sludge	0.25-2	0.25-2	weak *a*		*b*	0.73 0.93	weak *c*	high R_F		
2. Organic Sludge, Spiked	0.25-16	0.15-16	–		*c*	0.77 0.86	–		inconclusive	
3. EDC Spent Caustic	0.5-8	0.5-8	–		*b*	0.98	–		*a, b*	0.98
4. Herbicide Mfg. Acid	0.1-4	0.1-4	*a, b*	origin	–		*a, b*	origin	–	
5. Coke Plant Waste, DMSO	0.5-8	0.5-10	–		–		–		*c*	origin
6. Coke Plant Waste, EtOH	0.5-6	0.5-4	*c*	points of appl.	–		–		–	
7. Coke Plant Waste, untreated	0.5-10	0.5-10	–		*a*		–		–	
8. Herbicide Mfg. Acetone	0.25-4	0.5-4	–		– (*i*)		– (*i*)		– (*i*)	
9. Ink Pigment Sludge	0.5-4	0.5-16	–		–		–			
10. Paint Mfg. Emulsion	0.5-12	0.1-12	–		–		–		–	
11. Coal Gasification Tar	unknown	unknown	*a, b*		*a*		*c*	high R_F	–	
12. Organic Still Bottoms, EtOH Extract	0.1-8	0.1-8	–		–		–		–	
13. Municipal Sludge	1-16	0.5-12	*c*	origin	–		weak *a, c*	origin	–	
14. Pulp and Paper Waste	1-16	0.5-4	*b*	points of appl.	– (*i*)		*b* (*i*)	points of appl.	– (*i*)	
15. Oil Refining Waste	0.5-4	0.5-4	*c*	origin	toxic?		–		–	
16. Pharmaceutical Sludge, EtOH Extract	1-12	0.5-20	–		–		–		– (*i*)	

(*a*) 2-fold increase over spontaneous growth,
(*b*) Non-random cluster,
(*c*) Random cluster,
(*i*) Incomplete data; tested only once.

Table 4.3 – Summary of waste samples. (From [22], reprinted by permission of the authors and copyright holders).

SAMPLE	SOURCE*	IDENTITY	DESCRIPTION
1.	NBS	organic sludge	liquid sludge from an industrial solvent recovery process
2.	NBS	organic sludge, spiked	same as no. 10, but diluted and with 11 organics added
3.	Battelle	EDC spent caustic	wastes from ethylene dichloride production
4.	Battelle	herbicide mfg. acid	waste acid from a herbicide production process
5.	Battelle	coke oven waste, DMSO	coke plant waste extracted with DMSO
6.	Battelle	coke oven waste, EtOH	coke plant waste extracted with EtOH
7.	Battelle	coke oven waste, untreated	coke plant waste, no adulteration
8.	Battelle	herbicide acetone/ H_2O	water-soluble solvent from herbicide production
9.	Battelle	ink pigment	pourable black sludge
10.	Battelle	latex paint waste	polymeric emulsion from latex paint production
11.	Battelle	coal gasification tar	black, viscous tar produced from the gasification of coal
12.	Battelle	organic still bottoms	solid black sludge from tank bottoms
13.	Battelle	municipal sludge	chemically treated sludge; dark wet cake
14.	Battelle	pulp and paper waste	brown, wet fibrous cake
15.	Battelle	oil refining waste	dark oil with sediment
16.	Battelle	pharmaceutical sludge	gelatinous sludge from pharmaceuticals production

*NBS stands for the National Bureau of Standards, U.S. Department of Commerce, Washington, D.C. Battelle samples were obtained from Battelle Labs., Colombus, OH through Environmental Health Research Testing, Inc., Research Triangle Park, N.C.

The results reported as incomplete in Table 4.2 represent samples which were tested only once, the others being obtained from independently repeated experiments.

Figure 4.7 shows the results of one-dimensional TLC/*Salmonella* testing of organic sludge (sample 2) with tester strains TA 98 and TA 100 with and without S9. It reveals colonies of revertants at relative high R_F values for all systems.

Figure 4.8 shows the results of a one-dimensional TLC/*Salmonella* test of untreated herbicide manufacturing acid (sample 4). For this sample the response was positive only with S9 for the tester strains TA 98 and TA 100.

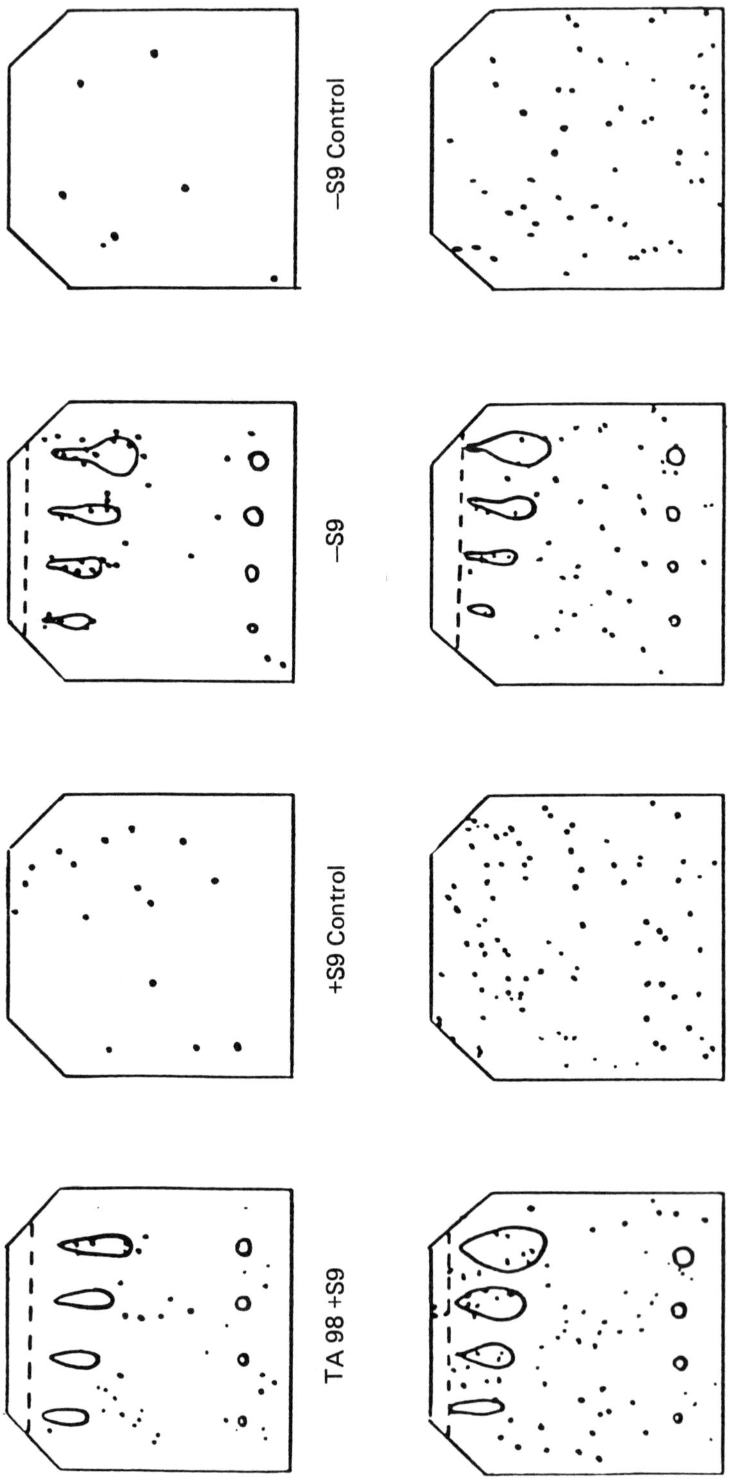

Fig. 4.7 – One-dimensional TLC/Ames test of organic sludge. (From [22], reprinted by permission of the authors and copyright holders).

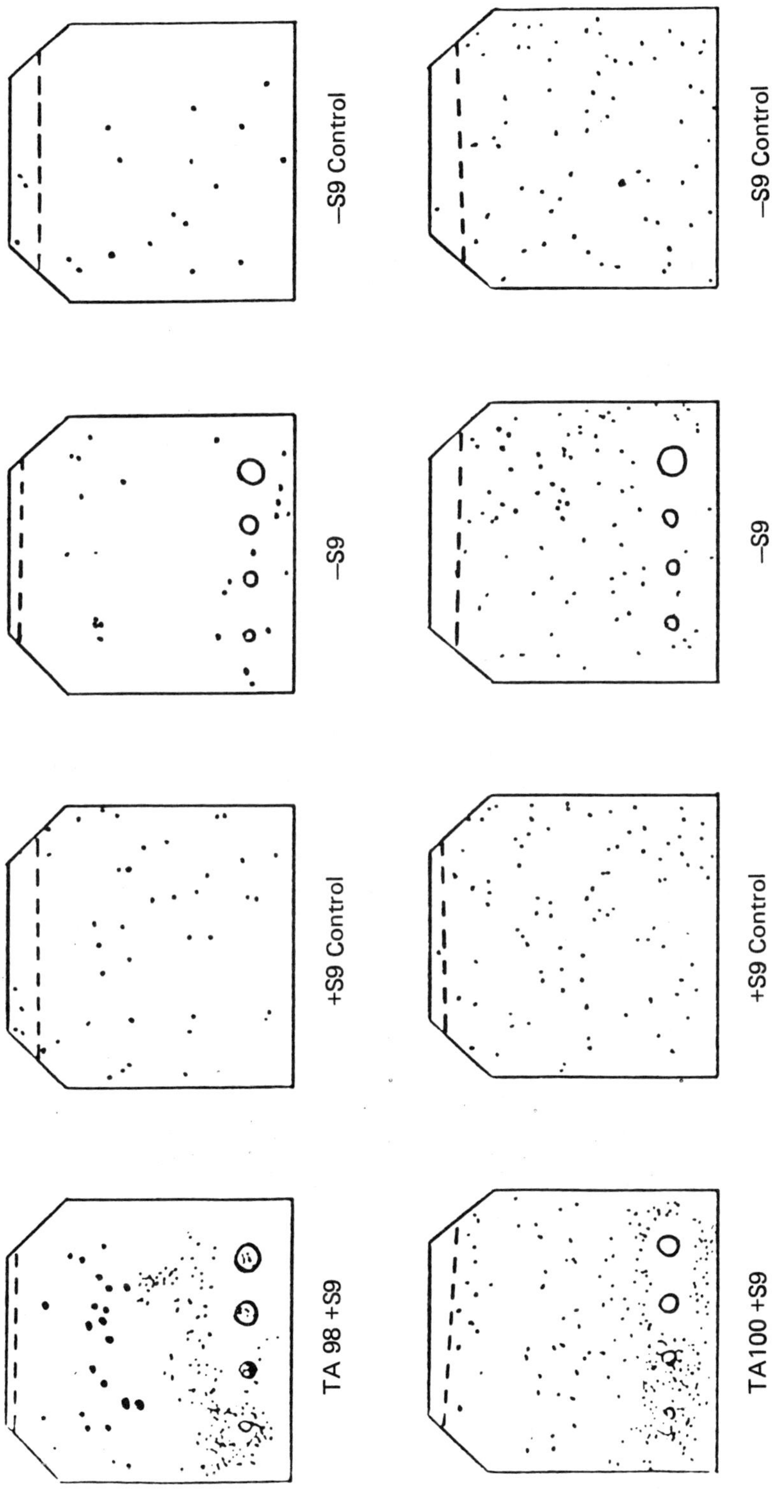

Fig. 4.8 – One-dimensional TLC/Ames test of herbicide manufacturing acid. (From [22], reprinted by permission of the authors and copyright holders).

The fact that samples (except for solids) can be tested without extraction is an important advantage of this technique over the standard Ames assay [22]. Another advantage that became evident during the course of the study of hazardous waste samples was that samples which were cytotoxic to *Salmonella* in the plate incorporation assay did not prove toxic with the TLC/*Salmonella* technique [23].

It was found in some cases that one-dimensional TLC gave revertants which disappeared in two-dimensional TLC. This might be due to the inherent diffusion of spots in two-dimensional TLC, which gives lower concentrations of potential mutagens for the *Salmonella* assay.

4.4 THE TLC/*SALMONELLA* ASSAY AS A SCREENING TECHNIQUE

Recently, considerable efforts have been put into attempts at characterization of environmental samples, by combining chromatographic fractionation of samples with subsequent bioassay. The biologically active fractions may then be analysed by various methods (for a review of such techniques, see Alfheim *et al.* [24]). A tremendous analytical effort is usually necessary to completely characterize an effluent. The cost therefore prohibits complete analysis of the large number of samples needed for environmental assessment. The TLC/*Salmonella* assay may, however represent an alternative which meets the requirements for a quick, simple and inexpensive test method.

In order to compare the separation abilities and the simplicity of TLC and HPLC for detection of mutagens, an extract of typewriter ribbon was fractionated into four fractions by HPLC and each fraction was tested by the conventional Ames test [6]. Each fraction, and the total extract, was also tested by the TLC/*Salmonella* assay. The results, given in Fig. 4.9, show that fraction 2 contained the mutagens from area I, which is probably due to nitropyrenes [9]. Fraction 4 contained the mutagens from area II. No mutagenic activity was obtained for fractions 1 and 3. The mutagen(s) in area III of the original extract could not be detected in any of the HPLC fractions. The R_F values suggest that these compounds are polar, indicating that they are not eluted from the HPLC column.

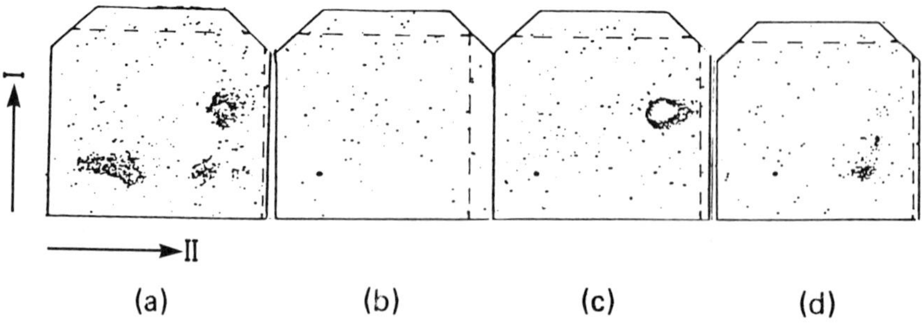

Fig. 4.9 – Comparison of TLC/Ames test with test of HPLC-separated samples. (a) is unfractionated, (b) is fraction 1 + 3, (c) is fraction 2 and (d) is fraction 4. Solvent systems as in [9].

These results show that TLC fractionation may be sufficient for separating the mutagens in a sample, and for polar compounds it offers advantages. An additional advantage is that all fractions can be tested for mutagenicity with only a single plate.

For preliminary testing of a large number of samples containing unknown compounds the TLC/*Salmonella* assay therefore represents a quick and simple method.

4.5 CONCLUSIONS

The TLC/*Salmonella* technique is still relatively new and the applications reported so far are limited. However, our experience and that of other scientists, suggests a few conclusions:

(a) the method is simple to perform and requires no advanced instruments or equipment;
(b) the method gives results quickly and can be applied to a large number of samples simultaneously;
(c) the method can be applied to extracts as well as to untreated samples (except solids);
(d) the method can be applied with various tester strains with or without metabolic activation;
(e) the separation of cytotoxic compounds from mutagens gives considerable advantages for their detection;
(f) the method still shows deficiencies, particularly in that it does not detect some recognized mutagens [benzo(*a*)pyrene and other PAH and chlorinated organic compounds].

REFERENCES

[1] M. D. Waters, S. S. Sandhu, J. L. Huisingh, L. Claxton and S. Nesnow, (eds.), *Short-term Bioassay in the Analysis of Complex Environmental Mixtures.* Vol. II, Plenum Press, New York, 1980.
[2] M. D. Waters, S. S. Sandhu, J. Lewtas, L. Claxton, N. Chernoff and S. Nesnow, (eds.), *Short-term Bioassay in the Analysis of Complex Environmental Mixtures,* Vol. III, Plenum Press, New York, 1982.
[3] A. Bjørseth, G. Eidså, J. Gether, L. Landmark and M. Møller, *Science,* 1982, **215,** 87.
[4] B. Commoner, in L. H. Keith, (ed.), *Identification & Analysis of Organic Pollutants in water,* Ann Arbor, Michigan, 1976. p. 49.
[5] D. C. Fenimore and C. M. Davis, *Anal. Chem.,* 1981, **53,** 252A.
[6] B. N. Ames, W. E. Durston, E. Yamasaki and F. D. Lee, *Proc. Natl. Acad. Sci. U.S.A.,* 1973, **70,** 2281.
[7] B. N. Ames, J. McCann and E. Yamasaki, *Mutation Res.,* 1975, **31,** 347.
[8] V. S. Houk, Presented at the First Annual Meeting of the Genetoxicity and Environmental Mutagen Society, Research Triangle Park, N.C. October 14, 1983.
[9] M. Møller, I. Alfheim, G. Löfroth and E. Agurell, *Mutation Res.,* 1983, **119,** 239.
[10] I. Alfheim, J. G. T. Bergström, D. Jenssen and M. Møller, *Environ. Health Perspect.,* 1983, **47,** 177.
[11] M. Møller, L. Landmark, A. Bjørseth and L. Renberg, *Chemosphere,* 1984, **13,** 813.
[12] A. Sundvall, L. Balk, C. Ekström, U. Rannug and L. Renberg, *Mutation Res.,* 1983, **113,** 309.

[13] A. Bjørseth, G. E. Carlberg and M. Møller, *Sci. Total Environ.*, 1979, **11**, 197.
[14] K. Pfister and E. Sjöström, *Svensk. Papperstidn.*, 1978, **81**, 195.
[15] P. Ander, K.-E. Eriksson, M.-C. Kolar, K. Kringstad, U. Rannug, and C. Ramel, *Svensk Papperstidn.*, 1977, **80**, 454.
[16] G. E. Carlberg, N. Gjøs, M. Møller, K. O. Gustavsen, G. Tveten and L. Renberg, *Sci. Total Environ.*, 1980, **15**, 3.
[17] J. Stockman, L. Strömberg and F. De Sousa, *Cellulose Chem. Technol.*, 1980, **14**, 517.
[18] M. Ahlberg, L. Berghem, G. Nordberg, S. A. Persson, L. Rudling and B. Steen, *Environ. Health Perspect.*, 1983, **47**, 85.
[19] J. Lewtas, *Environ. Health Perspect.*, 1983, **47**, 141.
[20] T. Ramdahl, I. Alfheim, S. Rustad and T. Olsen, *Chemosphere*, 1982, **11**, 601.
[21] T. Ramdahl and M. Møller, *Chemosphere*, 1983, **12**, 23.
[22] V. S. Houk and D. Claxton, Submitted to *Mutation Research.*
[23] V. S. Houk, Personal communication (1984).
[24] I. Alfheim, A. Bjørseth and M. Møller, *CRC Critical Reviews in Environmental Control*, 1984, **14**, 91.

5

Mutagens in food and drinks, and their carcinogenicity

M. NAGAO, K. WAKABAYASHI and **T. SUGIMURA,**
National Cancer Center Research Institute, 1-1,
Tsukiji 5-chome, Chuo-ku, Tokyo 104, Japan

Epidemiological studies have shown that the incidence of cancer in various organs of the human body is related to diet [1, 2]. Foods have been suspected to contain various types of mutagens/carcinogens. Application of the Ames method made it possible to detect carcinogens as mutagens in complex food materials. In this way many new mutagenic compounds were found in ordinary foods and beverages, such as hamburgers, grilled fish, vegetables and coffee. Soy sauce was also found to contain mutagen precursors that could be converted into nitroso compounds. There is no quantitative correlation between mutagenicity and carcinogenicity, but many strong mutagens found in foods have later been demonstrated to be carcinogenic to animals. Thus, information on the existence of mutagens in foods should be useful for preventing cancer.

5.1 COOKED FOOD

5.1.1 Mutagenicity of cooked foods

Cooking of some protein foods results in the production of mutagen(s) [3-6]. Grilled fish and fried hamburger cooked under ordinary conditions were found to be mutagenic to *S. typhimurium* TA98 with, but only with, S9 mix [7]. TA98 was more susceptible to the mutagens in these cooked foods than was TA100. Figure 5.1 shows that the mutagenicity of fish cooked over a naked gas flame and hamburger fried on a electric hot-plate increased with length of cooking. Usually fish is cooked for about 5 min and hamburger for about 15 min. With these cooking times, the mutagenicity of fish is 1000 times greater than that of hamburger. The difference could be due to the difference in cooking conditions, water content and/or the presence of precursors of mutagens.

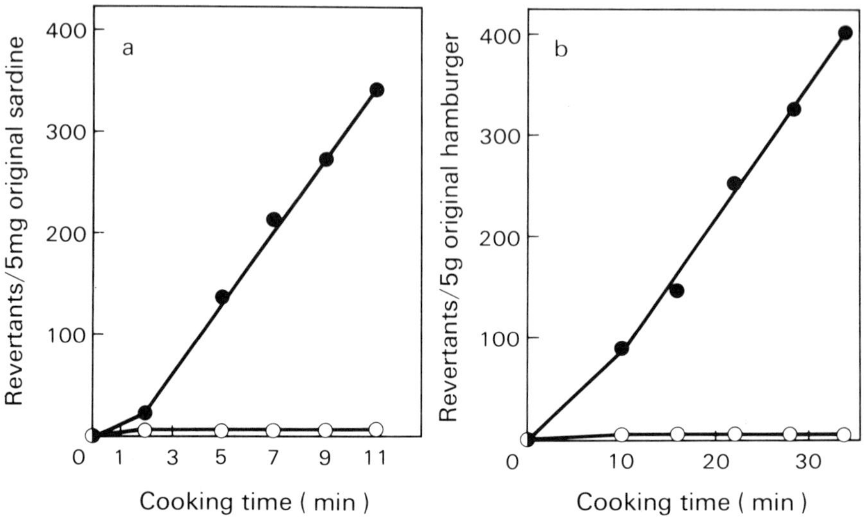

Fig. 5.1 – Mutagenicities of methanolic extracts of grilled sardine (*a*) and fried hamburger (*b*) to *S. typhimurium* TA98 with (●) or without (○) S9 mix after various time of cooking. [Reproduced by permission from T. Sugimura, in *Molecular Interrelations of Nutrition and Cancer,* M. S. Arnott, J. van Eys and Y.-M. Wang (eds.), pp. 3–24; copyright 1982, Raven Press, New York.]

Studies on the effects of heating various food components such as proteins, amino-acids, starch, nucleic acids and vegetable oil, showed that heated protein and amino-acids had strong mutagenicity to TA98 with S9 mix [8–10]. The heated products of starch, nucleic acid and vegetable oil showed little or no mutagenicity to TA98 with S9 mix. Pyrolysed starch had direct-acting mutagenicity on TA100.

5.1.2 Structures of mutagens isolated from heated materials

Fifteen mutagenic compounds were isolated from heated foods, amino-acids and proteins, and purified, and of these ten were newly identified compounds. Their structures are shown in Fig. 5.2, and their chemical names and abbreviations are given in Table 5.1.

The aminoimidazoquinoline derivatives IQ and MeIQ were purified as major mutagens from grilled sardine [11, 12], and an aminoimidazoquinoxaline derivative, MeIQx, was isolated from fried beef [13]. The imidazole parts of these mutagens could be derived from creatinine, and the remaining parts could be formed from the Maillard-reaction product 2-methylpyridine or 2,5-dimethylpyrazine produced from amino-acids and sugar [14]. MeIQx has been isolated as a major mutagen from a heated mixture of creatinine, glycine and glucose [15]. 7,8-DiMeIQx has also been isolated as a minor mutagen from this heated mixture [16].

Seven mutagenic compounds were isolated from the products formed on heating various amino-acids. These compounds were Trp-P-1, Trp-P-2, Glu-P-1,

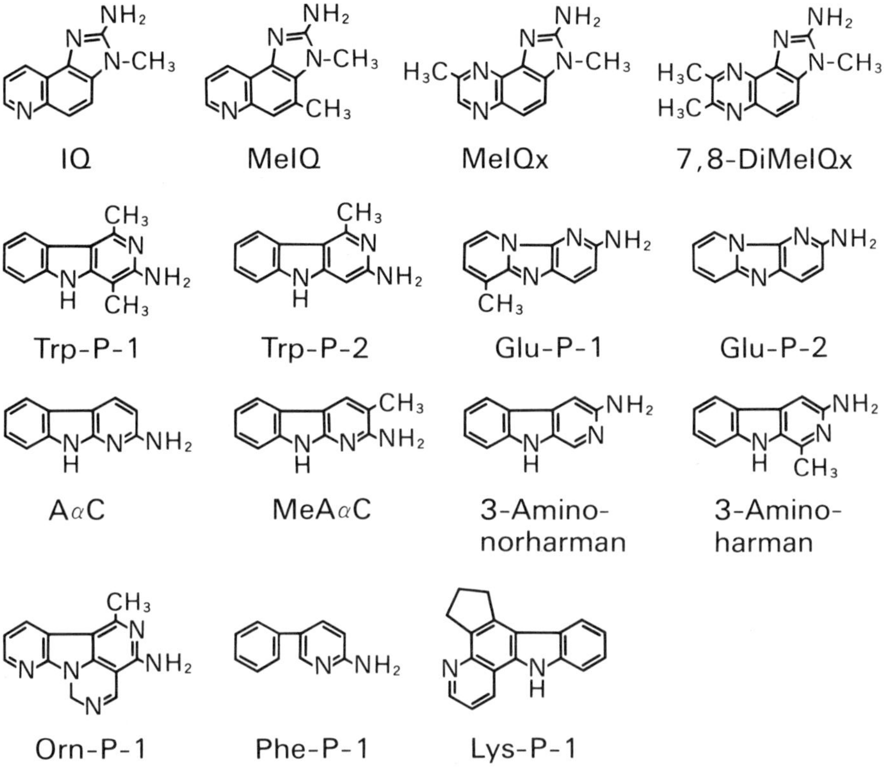

Fig. 5.2 – Structures of mutagens isolated from various heated materials.

Glu-P-2, Orn-P-1, Phe-P-1 and Lys-P-1 [17–20]. Two mutagenic compounds, AαC and MeAαC, were isolated from heated soybean globulin [21]. In addition, 3-aminonorharman and 3-aminoharman were isolated from a casein pyrolysate as substances causing sister-chromatid exchanges [22].

The structures of all mutagenic compounds described above were determined by X-ray crystallography or by comparison of their physical properties with those of synthetic materials. All except Lys-P-1 are heterocyclic amines.

5.1.3 Mutagenicities of pyrolysis products to bacteria and mammalian cells

The mutagenic activities to *S. typhimurium* TA98 and TA100 of mutagens isolated from various heated materials are given in Table 5.1 [7, 23, 24]. All mutagenic compounds required a microsomal metabolic activation system for appearance of mutagenicity. All heterocyclic amines were more mutagenic to TA98, a frameshift mutant, than to TA100, a base-pair change mutant. Conversely, Lys-P-1, which is a heterocyclic aromatic compound without an amino group, was more mutagenic to TA100 than to TA98. The mutagenicities of MeIQ, IQ, 7,8-DiMeIQx, MeIQx, Trp-P-2, Orn-P-1, Glu-P-1 and Trp-P-1 toward

Table 5.1 – Abbreviations, full names, sources and mutagenicities of pyrolysis products

Abbreviation	Full name	Source	Mutagenicity		
			S. typhimurium (Revertants/μg)[a]		Chinese hamster lung cells[b]
			TA98	TA100	(DT^r mutants/10^6 cells/μg)
IQ	2-Amino-3-methylimidazo-[4,5-*f*]quinoline	Grilled sardine	433×10^3	7.0×10^3	40
MeIQ	2-Amino-3,4-dimethylimidazo-[4,5-*f*]quinoline	Grilled sardine	661×10^3	30×10^3	38
MeIQx	2-Amino-3,8-dimethylimidazo-[4,5-*f*]quinoxaline	Fried beef	145×10^3	14×10^3	5.7
7,8-DiMeIQx	2-Amino-3,7,8-trimethylimidazo-[4,5-*f*]quinoxaline	Heated mixture of creatinine, glycine and glucose	163×10^3	9.9×10^3	
Trp-P-1	3-Amino-1,4-dimethyl-5*H*-pyrido[4,3-*b*]indole	Tryptophan pyrolysate	39×10^3	1.7×10^3	33
Trp-P-2	3-Amino-1-methyl-5*H*-pyrido[4,3-*b*]indole	Tryptophan pyrolysate	104×10^3	1.8×10^3	160
Glu-P-1	2-Amino-6-methyldipyrido-[1,2-*a*:3′,2′-*d*]imidazole	Glutamic acid pyrolysate	49×10^3	3.2×10^3	1.2
Glu-P-2	2-Aminodipyrido[1,2-*a*:3′,2′-*d*]-imidazole	Glutamic acid pyrolysate	1.9×10^3	1.2×10^3	0.3
Orn-P-1	4-Amino-6-methyl-1*H*-2,5,10,10b-tetraazafluoranthene	Ornithine pyrolysate	57×10^3		
Phe-P-1	2-Amino-5-phenylpyridine	Phenylalanine pyrolysate	41	23	
Lys-P-1	3,4-Cyclopentenopyrido-[3,2-*a*]carbazole	Lysine pyrolysate	90	100	
AαC	2-Amino-9*H*-pyrido[2,3-*b*]indole	Soybean globulin	300	20	20
MeAαC	2-Amino-3-methyl-9*H*-pyrido-[2,3-*b*]indole	Soybean globulin pyrolysate	200	120	
3-Aminonorharman	3-Amino-9*H*-pyrido[3,4-*b*]indole	Casein pyrolysate	0.4	0.3	
3-Aminoharman	3-Amino-1-methyl-9*H*-pyrido-[3,4-*b*]indole	Casein pyrolysate	0	0	

a: Amount of S9 per plate: IQ, MeIQ, MeIQ, MeIQx, 7,8-DiMeIQx, Trp-P-1, Trp-P-2, 10 μl; Glu-P-1, Glu-P-2, 30 μl; 3-aminonorharman, 3-aminoharman, 50 μl;
b: Amount of S9 per ml: Trp-P-2, 12 μl; the other compounds, 25 μl.

TA98 with S9 mix were extremely high and stronger than that of aflatoxin B_1 to TA98.

The mutagenicity of heterocyclic amines with S9 mix was also demonstrated with TA1538, but not with TA1535, neither of which has plasmid pKM101 [25, 26]. The mutagenicity to TA1538 was comparable with that to TA98.

Heterocyclic amines were mutagenic toward *Escherichia coli* WP2 *uvrA* and *E. coli* WP2 *uvrA*/pKM101 with S9 mix, the latter being the more susceptbile [26]. These compounds also induced prophage λ in lysogenic *E. coli* K12, strain GY5027 [26].

The mutagenicity of eight heterocyclic amines was examined by using Chinese hamster lung cells *in vitro* with diphtheria toxin resistance as a selective marker [27]. The mutation frequencies of these cells induced by heterocyclic amines are shown in Table 5.1. All the heterocyclic amines tested except Trp-P-1 required S9 mix to exhibit mutagenicity. Trp-P-1 showed the same mutagenicity with and without S9 mix. The mutagenic activities of all the heterocyclic amines were lower than that of 4-nitroquinoline-*N*-oxide and comparable with those of benzo[*a*]pyrene and *N,N*-dimethylnitrosamine [27].

Trp-P-1, Trp-P-2, Glu-P-1, AαC, 3-aminonorharman and 3-aminoharman induced sister-chromatid exchanges in human lymphoblastoid cells NL-3 [22, 28]. Trp-P-1, Trp-P-2, Glu-P-1, Glu-P-2, AαC and MeAαC induced chromosome aberrations in a Chinese hamster cell line of lung fibroblast origin [29].

5.1.4 Modification of mutagenicity of heterocyclic amines

Heterocyclic amines, such as Trp-P-1, Glu-P-1 and AαC, were inactivated by treatment with nitrite under acidic conditions [30–32]. The non-mutagenic compounds were shown to be hydroxy derivatives formed by deamination. In contrast, aminoimidazoquinoline and aminoimidazoquinoxaline compounds, such as IQ, MeIQ and MeIQx, were completely resistant to nitrite treatment [32].

All heterocyclic amines were inactivated by treatment with chlorinated tap water or hypochlorite [32, 33]. This reaction proceeded very rapidly: samples of about 200 μg of the heterocyclic amines were converted into non-mutagens within a few minutes at room temperature on treatment with 150 ml of tap water containing 1.5 ppm of residual chlorine. The product formed from Glu-P-1 with hypochlorite was shown to be the azo-dimer of Glu-P-1 [25]. Based on their responses to nitrite, heterocyclic amines can be classified into two types, represented by IQ and Trp-P-1. The contributions of the two types of mutagents to the total mutagenicity of the basic function of the pyrolysed mixture can easily be determined [32].

Heterocyclic amines were inactivated by peroxidases, such as myeloperoxidase, lactoperoxidase and horseradish peroxidase, in the presence of hydrogen peroxide [34].

The mutagenicity of heterocyclic amines was also inhibited by haemin, and oleic and linoleic acids [35, 36]. Vegetable fibres reduced the mutagenicities of heterocyclic amines by adsorbing these compounds [37]. Cotton fibre also adsorbed heterocyclic amines to some extent [38]. Blue cotton, which is cotton fibre with covalently bound trisulpho-copper-phthalocyanine, adsorbed compounds with a polycyclic planar structure very efficiently, and the compounds adsorbed could be eluted quantitatively with methanol–ammonia [38]. Therefore, blue cotton is very useful for concentrating heterocyclic amines from crude materials such as heated foods, urine, cigarette-smoke condensate and river water.

5.1.5 Carcinogenicity of heterocyclic amines

IQ, Trp-P-1, Trp-P-2, Glu-P-1, Glu-P-2, AαC and MeAαC were found to be carcinogenic to CDF_1 mice when given orally; IQ induced tumours of the liver, fore-stomach and lung [39]. Trp-P-1, Trp-P-2, Glu-P-1, Glu-P-2, AαC and MeAαC induced liver tumours, and the last four compounds also induced tumours in the brown adipose tissue [40, 41]. Female mice were more susceptible than males to induction of liver tumours by these heterocyclic amines.

IQ, Glu-P-1, Glu-P-2 and Trp-P-2 were demonstrated to be carcinogenic to rats on oral administration. IQ, Glu-P-1 and Glu-P-2 were multipotent carcinogens in F344 rats [42, 43]. They induced tumours in the liver, small and large intestines, Zymbal gland, clitoral gland and brain. Trp-P-2 induced liver tumours in female ACI rats [44].

5.1.6 Amounts of heterocyclic amines in heated foods

Although no standard method for quantitative analysis has yet been established, the distribution of heterocyclic amines in various heated foods has been determined. For determination of heterocyclic amines in heated foods, the following procedures for partial purification of heterocyclic amines are useful: extraction with methanol or dilute hydrochloric acid, adsorption on blue cotton, separation by acid–base partition, Sephadex LH-20 column chromatography and silica-gel column chromatography. The partially purified samples obtained by these procedures can then be analysed by gas chromatography/mass spectrometry with multiple-ion detection, and by high-pressure liquid chromatography on an octadecylsilyl column with monitoring by ultraviolet-absorption or fluorescence. The amounts of heterocyclic amines detected in various heated foods are given in Table 5.2 [24, 44–51].

The content of *N,N*-dimethylnitrosamine in salt-dried fish was also found to be increased by grilling [52]. 1-Nitropyrene, a strong frameshift mutagen, was also found in grilled chicken [53].

5.2 VEGETABLES

5.2.1 Flavonoids

Flavonoids are common in vegetables, and their intake by humans may be as much as 1 g per day. The mutagenicity of flavonoids was observed by Wang *et al.* [54], Bjeldanes and Chang [55], Sugimura *et al.* [56], and Hardigree and Epler [57]. Later many flavonoids were subjected to the microbial mutagenicity test [58–63]. The skeleton of flavone is shown in Fig. 5.3. The structural requirement for mutagenicity is a hydroxy group at position 3 [57] or a hydroxy

Table 5.2 – Amount of heterocyclic amines in various heated foods (ng/g of pyrolysate)

	Sun-dried sardine	Beef	Beef[a]	Hamburger[a]	Food-grade beef extract	Sun-dried cuttle fish	Chicken	Chinese mushroom	Onion
IQ	158			0.02					
MeIQ	72								
MeIQx			2.4	1.0	28				
Trp-P-1	13.3	53[b]							
Trp-P-2	13.1								
Glu-P-2						280[b]			
AαC		650.8					180.4	47.2	1.5
MeAαC		63.5					15.1	5.4	

a: Fried on electric hot-plate. Other foods were heated over a naked gas flame.

b: Per g of raw material.

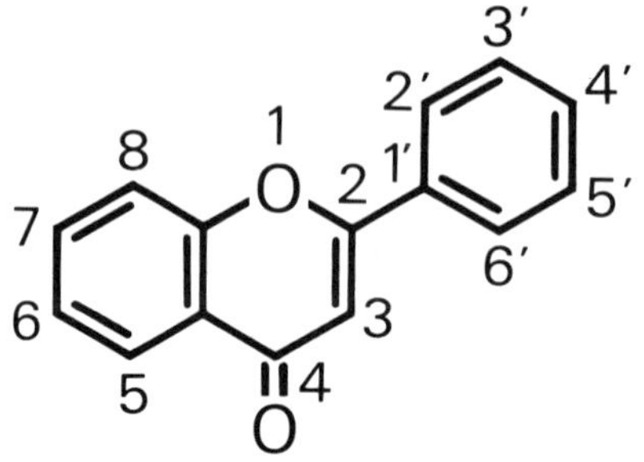

Flavone

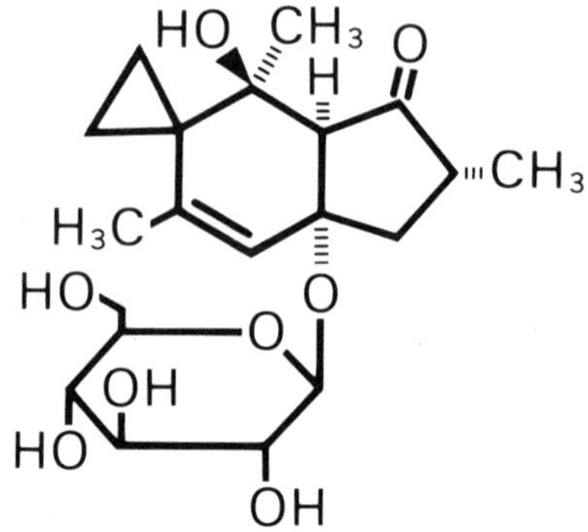

Aquilide A

Fig. 5.3 – Structures of flavone and aquilide A.

or methoxy group at position 8 [61, 63]. 3-Hydroxyflavones (flavonols) include quercetin, kaempferol, rhamnetin and isorhamnetin, and 8-hydroxy- or methoxy-flavones include wogonin and norwogonin. Among flavonoids, these compounds have comparatively strong mutagenicity. The mutagenicities of quercetin and wogonin with S9 mix were 19 revertants/nmole on TA98 and 9.8 revertants/nmole on TA100, respectively. 3-*O*-Glycosides of flavonols exhibit mutagenicity after acid hydrolysis or treatment with glycosidases, such as fecalase [64], cecalase [65] or crude hesperidinase [61].

The genotoxicity and carcinogenicity of the flavonoid quercetin have been studied extensively. Quercetin induced mutation at the thymidine kinase locus in Chinese hamster ovary fibroblasts [66] and L5178Y mouse lymphoma cells [67–69] *in vitro*. However, consistent results were not obtained for induction of a micronucleus in mouse bone marrow cells [70–73]. Conflicting results have also been obtained on the carcinogenicity of quercetin, which was found to be not carcinogenic in rats [74–76], mice [77–80] or hamsters [81], but carcinogenicity has been found in rats [82], with a high incidence of intestinal tumours being induced.

Mutagenic flavonoids are extremely common, and flavonoids have been isolated as major mutagens in methanol extracts of various foods, pickles [83], spices [84], tea infusions [85] and citrus fruit juice [86].

5.2.2 Pickles

A methanol extract of Japanese pickles consisting of a mixture of 9 kinds of vegetables was shown to be mutagenic to TA98 with S9 mix. Its specific activity was 3.6 revertants/mg (wet weight) of these pickles. The active principles were purified; the major one was identified as kaempferol and a minor one as isorhamnetin [83]. Methanol extracts of 8 of the 9 vegetables showed mutagenicity to TA98 with S9 mix, their specific activities each being between 1.8 and 5.7 revertants/mg (wet weight) [83].

A pickled mixture of vegetables produced in Linxian County in China has been suspected to be a cause of the high incidence of human oesophageal cancer in that county [87]. Methanol extraction and successive methanol–chloroform (1:1) extraction of these pickles yielded material with mutagenicity to TA100 and TA98 with S9 mix. The mutagenicity of these Chinese pickles was about 6 times that of the Japanese pickles mentioned above [83]. Although identification of the major mutagen(s) has not yet been achieved, one minor mutagen was identified as Roussin's red methyl ester, which is only weakly mutagenic to TA100 with S9 mix, and is present at about 1 ppm in the Chinese pickles [88]. The mouldy corn bread which is frequently found in pickles, was found to contain a new mutagenic nitrosamine, *N*-3-methylbutyl-*N*-1-methylacetonyl-nitrosamine [89].

5.2.3 Spices

Methanol extracts of dill weed, sumac, thyme, pot-pourri, oregano, bay leaves and marjoram showed mutagenicity to TA100 or TA98 with S9 mix. The specific activities of the original spices were less than 10 revertants/mg. The contribution of spices to the overall mutagenicity of ordinary foods is probably low, because very small amounts of spices are actually used and their specific mutagenic activities are low. The major mutagenic principles were found to be quercetin 3-sulphate and isorhamnetin 3-sulphate in dill weed [90], and quercetin in sumac [84].

5.2.4 Bracken

Bracken is known to induce intestinal and bladder tumours in experimental animals [91, 92]. The amounts of quercetin and kaempferol present as glycosides in bracken were estimated to be 0.57 and 1.1 g, respectively, per kg of the dried material [82]. A methanol–light petroleum extract of freeze-dried bracken has been demonstrated to be mutagenic to TA98 and TA100 without S9 mix after alkaline treatment. Aquilide A (Fig. 5.3) was isolated by Van der Hoeven *et al.* [93] from the methanol–light petroleum extract as a mutagenic principle showing mutagenicity only after alkaline treatment. This compound, named ptaquiloside, was also isolated from bracken as a carcinogenic principle by Hirono *et al.* [94]. It induced mammary cancer in female CD rats when given intragastrically, as bracken did when given orally.

5.2.5 Other mutagens

Plants contain various other mutagenic carcinogens. The flower stalks of ***Petasites japonicus*** contain a pyrrolizidine alkaloid, petasitenine (synonym, fukinotoxin) which is hepatocarcinogenic [95] and mutagenic to TA100 with S9 mix [96]. The green vegetable *Symphytum officinale,* commonly called comfrey or Russian comfrey, contains the pyrrolizidine alkaloids symphytine and echimidine. Symphytine was found to be mutagenic [97] and carcinogenic [98]. The mutagenicity of pyrrolizine alkaloids was weak, but dose-dependent effects were clearly observed by the Ames method with the modificaton of preincubation [96]. Heliotrine, lasiocarpine, clivorine, ligularidine and senkirkine were found to be mutagenic [96].

Various mutagenic and carcinogenic hydrazine derivatives were reported to be present in various mushrooms: *Gyromitra esculenta* contains carcinogenic gyromitrin (synonym, acetaldehyde methylformylhydrazone) at 50 mg/kg [99, 100]: *Agaricus bisporus* contains this compound at 0.4 mg/kg, and also 4-(hydroxymethyl)benzene diazonium ion, which is a degradation product of agaritin with mutagenicity to TA1535 and TA1537 [101] and carcinogenicity in mice [102].

5.3 PRECURSORS OF MUTAGENS THAT BECOME MUTAGENIC AFTER NITRITE TREATMENT

Of exogenously supplied or endogenously formed nitrate, about 20% is excreted in the saliva in humans and it is known to be reduced to nitrite by oral flora. Nitrite reacts with secondary amines, amides or phenols and produces mutagenic and carcinogenic compounds under the acidic conditions in the stomach [103].

Known mutagenic and carcinogenic nitroso-compounds have been determined and the necessary improvements in food processes have been made [104].

Precursors of mutagens that become mutagenic after nitrite treatment were recently found in soy sauce, fish and fava beans by the *Salmonella* mutagenicity test. Indole compounds seem to be significant as precursors of mutagens and are possibly hazardous to humans.

5.3.1 Soy sauce

Soy sauce contains direct-acting mutagenicity to TA100 after nitrite treatment [105]. Volumes of 1 ml of various soy sauces produced in Japan induced 2700–25200 revertants of TA100 without S9 mix after nitrite treatment. When fermented soybeans (*natto*) and bean paste, which is also a fermented product of soybean, were treated with nitrite, only the latter was found to show mutagenicity. Soybeans and their products such as bean curd, bean milk and boiled green soybeans were not mutagenic after nitrite treatment [105].

Three precursors of mutagens that become mutagenic on nitrite treatment were isolated from soy sauce and identified as (–)-(1*S*,3*S*)-1-methyl-1,2,3,4-tetrahydro-β-carboline-3-carboxylic acid [(–)-(1*S*,3*S*)-MTCA], its stereoisomer (–)-(1*R*,3*S*)-MTCA and tyramine [105, 106] (Fig. 5.4). The mutagen produced from (–)-(1*S*,3*S*)-MTCA and nitrite was a minor product, and its structure has not yet been elucidated. A major product was the *N*-2 nitroso-substituted compound, (–)-(1*S*,3*S*)-1-methyl-2-nitroso-1,2,3,4-tetrahydro-β-carboline-3-carboxylic acid, but it was not mutagenic. A direct acting mugaten was isolated as a major product from the reaction mixture of tyramine and nitrite. Its structure was determined to be 4-(2-aminoethyl)-6-diazo-2,4-cyclohexadienone [106]. One μg of this mutagen induced 112 revertants of TA100.

The amounts of MTCAs and tyramine in various soy sauces produced in Japan were 82–678 and 17–2250 μg/ml, respectively. The amounts of these

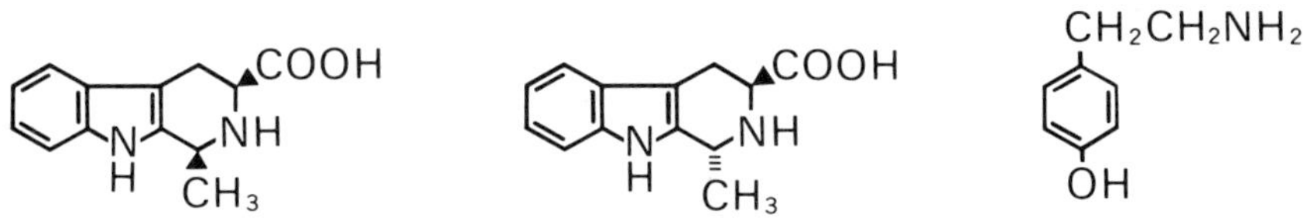

(−)-(1*S*, 3*S*)-MTCA (−)-(1*R*,3*S*)-MTCA Tyramine

Fig. 5.4 – Structures of precursors of mutagens isolated from soy sauce.

precursors in most soy sauces produced in the USA, of which the mutagenicities after nitrite treatment were weaker, were far less or not detectable. MTCA and tyramine seem to be formed in soy sauce during the fermentation and brewing processes; MTCA is formed by condensation of L-tryptophan and acetaldehyde, and tyramine is formed by decarboxylation of tyrosine. Soy sauce is widely used as a seasoning in Southeast Asia. In Japan, the daily consumption of soy sauce is about 30 ml *per capita*. Therefore, the daily ingestions of MTCAs and tyramine in soy sauce are at most about 20 and 30 mg *per capita*, respectively.

5.3.2 Fava bean

Fava beans, which are a major constituent of the diet of the inhabitants of Colombia, were found to be mutagenic to *S. typhimurium* TM677 without S9 mix after nitrite treatment [107]. Recently, 4-chloro-6-methoxyindole was isolated as a mutagen precursor [108].

5.3.3 Pepper

Nitrite-treated pepper showed direct-acting mutagenicity to TA98 and TA100 [109]. A major component of pepper, piperine, also showed direct-acting mutagenicity after nitrite treatment. 6-Nitropiperonal was identified as one mutagen produced from piperine with nitrite [110].

5.3.4 Fish

Specimens of the mackerel-pike ("sanma" in Japanese) obtained in New York were found to show mutagenicity to TA100 and TA135 without S9 mix after nitrite treatment [111]. On repeated gastric intubation, an extract of nitrite-treated fish induced tumours of the glandular stomach of rats [112]. The structure of the mutagen or its precursor is not yet clarified.

5.3.5 Chinese cabbage

Chinese cabbage is very commonly eaten in Japan. Nitrite-treated Chinese cabbage showed direct-acting mutagenicity to TA100, and indoleacetonitrile was isolated from it as a precursor of mutagens [113].

5.4 MUTAGENS IN BEVERAGES

5.4.1 Coffee

Coffee is a common beverage throughout the world, and about three-five cups of coffee is a common daily consumption. The mutagenicities of coffee to *S. typhimurium* and *E. coli* were demonstrated by the preincubation method [114-116]. One small cup of coffee (1.5 g of instant coffee in 100 ml of water) contains mutagens inducing about 5×10^4 revertants of *S. typhimurium* TA100. The mutagens were equally mutagenic to *E. coli* WP2/*uvrA*/pKM101 but less mutagenic to *S. typhimurium* TA98. Freshly brewed coffee (10 g of beans in 100 ml) also shows similar activity. Both fresh and instant coffee induced diphtheria toxin resistance mutation (DT^r) of Chinese hamster lung cells *in vitro* [117]. Instant coffee at 8 mg/ml induced 100 DT^r cells per 2.5×10^5 survivors. As mutagenic principles in coffee, carbonyl compounds, glyoxal and biacetyl [118] and methylglyoxal [119] have been identified. The amounts of these diketones in one brand of instant coffee and their mutagenic activity are shown in Table 5.3. Among these diketones, methylglyoxal is the commonest and has the highest mutagenicity. However, methylglyoxal cannot account for the major mutagenicity of instant coffee and freshly brewed coffee [116], because most of the mutagenicity of these coffees is suppressed by catalase, whereas that of methylglyoxal is not. Actually, both fresh and instant coffee contain hydrogen peroxide; instant coffee (at 15 mg/ml) contains 3 ppm soon after preparation. However, hydrogen peroxide itself cannot account for the mutagenicity of coffee. It seems possible that some precursor(s) of a mutagen may become active on reaction with hydrogen peroxide [116], and the active compound(s) may be unstable. A boiling-water extract of green coffee beans showed no mutagenicity and contained far less hydrogen peroxide. Mutagens in coffee seem to be produced during roasting of the coffee beans.

Table 5.3 – Mutagenicities of diketones and their concentrations in instant coffee

Diketone	Structure	Mutagenicity[a] (revertants/mg)	Concentration (μg/g)
Glyoxal	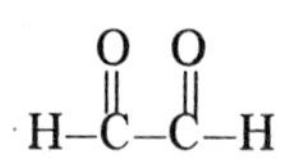	9×10^3	915
Methylglyoxal	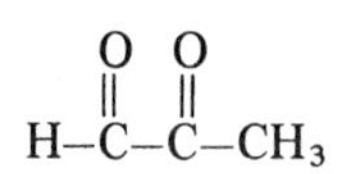	100×10^3	210
Biacetyl	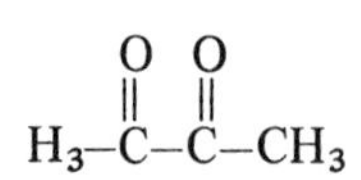	360	80

a TA100 was used without S9 mix.

5.4.2 Alcoholic beverages

Acetaldehyde, which is formed metabolically from ethanol in the human body, is known to induce sister-chromatid exchanges in Chinese hamster cells and human cells *in vitro* [120].

Evaporated residues of whisky, brandy and apple brandy have been reported to be mutagenic to *S. typhimurium* [121, 122]. The mutagenicity of those produced in western countries and Japan were shown with TA100 without S9 mix and were reduced by adding S9 mix. Since unbarrelled alcoholic spirits showed no mutagenicity, the mutagens in brewed alcoholic beverages may be extracted from the burnt surface of barrels in which the distilled alcohols are stored for several years. Storage may also be related to the production of mutagens. The mutagenicity of evaporated residues of whisky was not affected by catalase, unlike the mutagen(s) in coffee. The main mutagens in whisky and brandies have not yet been identified.

Some Chinese alcoholic spirits showed mutagenicity to TA98 and TA100 and some required S9 mix for mutagenicity. Differences in the ingredients and methods of brewing must be the cause of the differences in properties between the alcoholic products made in Western countries and Japan or China [123].

5.5 DISCUSSION

Short-term tests with the microbial mutation system were found to be effective in the search for unidentified environmental carcinogens as mutagens. Advances in quantitative and qualitative analytical methods and in isolation and purification methods, and the revolutionary progress in methods for determination of chemical structures, have made the identification of new mutagens possible. Synthesis of newly-found mutagens then allows studies on the carcinogenic potencies of these compounds in experimental animals.

However, mutagens are not necessarily carcinogens, as indicated by the fact that the mutagen quercetin is not carcinogenic in long-term animal experiments, most studies on the carcinogenicity of quercetin have given negative results, although one positive result has been reported. Therefore the detection of mutagenicity alone in a certain food is not sufficient for any recommendation to be made to the general public. The *in vivo* metabolism of mutagens/carcinogens is an important factor in determining their potency as carcinogens. The balance between metabolic activation and detoxificaion is as important as that between the absorption and excretion of mutagens/carcinogens and their metabolites.

Recent studies on the mechanisms of activation of oncogenes in humans and experimental cancers show that point mutations, translocations and gene-amplifications are important in carcinogenesis. Environmental mutagens probably act in carcinogenesis by influencing these processes. However, carcinogenesis is a multistep process, and the various steps have not yet been fully elucidated at the molecular level. Thus the susceptibilities of animals to mutagens/carcinogens may be related to events not only in the initiation step, but also in the promotion step.

For estimation of the risks of environmental mutagens to humans further extensive basic studies are necessary, including dose-dependence studies and also two-step carcinogenesis experiments with use of environmental initiators and promoters. Thus at present it is premature to make any definite statements about the risks to humans of mutagens/carcinogens in foods.

REFERENCES

[1] R. Doll, *Nature (London)*, 1977, **265**, 589.
[2] E. L. Wynder and G. B. Gori, *J. Natl. Cancer Inst.*, 1977, **58**, 825.
[3] M. Nagao, M. Honda, Y. Seino, T. Yahagi and T. Sugimura, *Cancer Lett.*, 1977, **2**, 221.
[4] T. Sugimura, M. Nagao, T. Kawachi, M. Honda, T. Yahagi, Y. Seino, S. Sato, N. Matsukura, T. Matsushima, A. Shirai, M. Sawamura and H. Matsumoto, in *Origins of Human Cancer, Book B,* H. H. Hiatt, J. D. Watson and J. A. Winsten (eds.), Cold Spring Harbor Laboratories, Cold Spring Harbor, New York, 1977, p. 156.
[5] B. Commoner, A. J. Vithayathil, P. Dolora, S. Nair, P. Madyastha and G. C. Cuca, *Science,* 1978, **201**, 913.
[6] N. E. Spingarn and J. H. Weisburger, *Cancer Lett.*, 1979, **7**, 259.
[7] T. Sugimura, in *Molecular Interrelations of Nutrition and Cancer,* M. S. Arnott, J. van Eys and Y.-M. Wang (eds.), Raven Press, New York, 1982, p. 3.
[8] M. Nagao, M. Honda, Y. Seino, T. Yahagi, T. Kawachi and T. Sugimura, *Cancer Lett.*, 1977, **2**, 335.
[9] M. Nagao, T. Yahagi, T. Kawachi, Y. Seino, M. Honda, N. Matsukura, T. Sugimura, K. Wakabayashi, K. Tsuji and T. Kosuge, in *Progress in Genetic Toxicology,* D. Scott, B. A. Bridges and F. H. Sobels (eds.), Elsevier/North-Holland, Amsterdam, 1977, p. 259.
[10] T. Kosuge, K. Tsuji, K. Wakabayashi, T. Okamoto, K. Shudo, Y. Iitaka, A. Itai, T. Sugimura, T. Kawachi, M. Nagao, T. Yahagi and Y. Seino, *Chem. Pharm. Bull.*, 1978, **26**, 611.
[11] H. Kasai, Z. Yamaizumi, S. Nishimura, K. Wakabayashi, M. Nagao, T. Sugimura, N. E. Spingarn, J. H. Weisburger, S. Yokoyama and T. Miyazawa, *J. Chem. Soc. Perkin I,* 1981, 2290.
[12] H. Kasai, Z. Yamaizumi, K. Wakabayashi, M. Nagao, T. Sugimura, S. Yokoyama, T. Miyazawa and S. Nishimura, *Chem. Lett.*, 1980, 1391.
[13] H. Kasai, Z. Yamaizumi, T. Shiomi, S. Yokoyama, T. Miyazawa, K. Wakabayashi, M. Nagao, T. Sugimura and S. Nishimura, *Chem. Lett.*, 1981, 485.
[14] M. Jägerstad, A. Laser Peuterswärd, R. Öste and A. Dahlqvist, *The Maillard Reaction in Foods and Nutrition,* in G. R. Waller and M. S. Feather (eds.), American Chemical Society, Washington, D.C., 1983, p. 507.
[15] Jägerstad, K. Olsson, S. Grivas, C. Negishi, K. Wakabayashi, M. Tsuda, S. Sato and T. Sugimura, *Mutation Res.*, 1984, **126**, 239.
[16] C. Negishi, K. Wakabayashi, M. Tsuda, S. Sato, T. Sugimura, H. Saito, M. Maeda and M. Jägerstad, *Mutation Res.*, 1984, **140**, 55.
[17] T. Sugimura, T. Kawachi, M. Nagao, T. Yahagi, Y. Seino, T. Okamoto, K. Shudo, T. Kosuge, K. Tsuji, K. Wakabayashi, Y. Iitaka and A. Itali, *Proc. Japan Acad.*, 1977, **53**, 58.
[18] T. Yamamoto, K. Tsuji, T. Kosuge, T. Okamoto, K. Shudo, K. Takeda, Y. Iitaka, K. Yamaguchi, Y. Seino, T. Yahagi, M. Nagao and T. Sugimura, *Proc. Japan Acad.*, 1978, **54**, 248.
[19] M. Yokota, K. Narita, T. Kosuge, K. Wakabayashi, M. Nagao, T. Sugimura, K. Yamaguchi, K. Shudo, Y. Iitaka and T. Okamoto, *Chem. Pharm. Bull.*, 1981, **29**, 1473.
[20] K. Wakabayashi, K. Tsuji, T. Kosuge, K. Takeda, K. Yamaguchi, K. Shudo, Y. Iitaka, T. Okamoto, T. Yahagi, M. Nagao and T. Sugimura, *Proc. Japan Acad.*, 1978, **54**, 569.
[21] D. Yoshida, T. Matsumoto, R. Yoshimura and T. Matsuzaki, *Biochem. Biophys. Res. Commun.*, 1978, **83**, 915.
[22] M. Tada, H. Saeki and A. Oikawa, *Mutation Res.*, 1983, **121**, 81.
[23] T. Sugimura, *Cancer,* 1982, **49**, 1970.
[24] T. Sugimura and S. Sato, *Cancer Res. (Suppl.),* 1983, **43**, 2415s.

[25] T. Sugimura, S. Sato and S. Takayama, in *Environmental Aspects of Cancer: The Role of Macro and Micro Components of Foods,* E. L. Wynder, G. A. Leveille, J. H. Weisburger and G. E. Livingston (eds.), Food & Nutrition Press, Westport, Connecticut, 1982, p. 167.
[26] M. Nagao, S. Sato and T. Sugimura, in *The Maillard Reaction in Foods and Nutrition,* G. R. Waller and M. S. Feather (eds.), American Chemical Society, Washington, D.C., 1983, p. 521.
[27] M. Nakayasu, F. Nakasato, H. Sakamoto, M. Terada and T. Sugimura, *Mutation Res.,* 1983, **118,** 91.
[28] H. Tohda, A. Oikawa, T. Kawachi and T. Sugimura, *Mutation Res.,* 1980, **77,** 65.
[29] M. Ishidate, Jr., T. Sofuni and K. Yoshikawa, *GANN Monogr. Cancer Res.,* 1981, **27,** 95.
[30] M. Tsuda, Y. Takahashi, M. Nagao, T. Hirayama and T. Sugimura, *Mutation Res.,* 1980, **78,** 331.
[31] M. Tsuda, M. Nagao, T. Hirayama and T. Sugimura, *Mutation Res.,* 1981, **83,** 61.
[32] T. Sugimura and S. Sato, in *Developments in the Science and Practice of Toxicology,* A. W. Hayes, R. C. Schnell and T. S. Miya (eds.), Elsevier, Amsterdam, 1983, p. 115.
[33] M. Tsuda, K. Wakabayashi, T. Hirayama and T. Sugimura, *Mutation Res.,* 1983, **119,** 27.
[34] M. Yamada. M. Tsuda, M. Nagao, M. Mori and T. Sugimura, *Biochem. Biophys. Res. Commun.,* 1979, **90,** 769.
[35] S. Arimoto, Y. Ohara, T. Namba, T. Negishi and H. Hayatsu, *Biochem. Biophys. Res. Commun.,* 1980, **92,** 662.
[36] H. Hayatsu, S. Arimoto, K. Togawa and M. Makita, *Mutation Res.,* 1981, **81,** 287.
[37] T. Kada, M. Kato, K. Aikawa and S. Kiriyama, *Mutation Res.,* 1984, **141,** 149.
[38] H. Hayatsu, T.Oka, A. Wakata, Y. Ohara, T. Hayatsu, H. Kobayashi and S. Arimoto, *Mutation Res.,* 1983, **119,** 233.
[39] H. Ohgaki, K. Kusama, N. Matsukura, K. Morine, H. Hasegawa, S. Sato, S. Takayama and T. Sugimura, *Carcinogenesis,* 1984, **5,** 921.
[40] H. Ohgaki, N. Matsukura, K. Morino, T. Kawachi, T. Sugimura and S. Takayama, *Carcinogenesis,* 1984, **5,** 815.
[41] N. Matsukura, T. Kawachi, K. Morino, H. Ohgaki, T. Sugimura and S. Takayama, *Science,* 1981, **213,** 346.
[42] S. Takayama, M. Masuda, M. Mogami, H. Ohgaki, S. Sato and T. Sugimura, *Gann,* 1984, **75,** 207.
[43] S. Takayama, Y. Nakatsuru, M. Masuda, H. Ohgaki, S. Sato and T. Sugimura, *Gann,* 1984, **75,** 467.
[44] S. Hosaka, T. Matsushima, I. Hirono and T. Sugimura, *Cancer Lett.,* 1981, **13,** 23.
[45] T. Sugimura, M. Nagao and K. Wakabayashi, in *Environmental Carcinogens – Selected Methods of Analysis, Vol. 4,* H. Egan, L. Fishbein, M. Castegnaro, I. K. O'Neill, H. Bartsch and W. Davies (eds.), Agency for Research on Cancer, Lyon, 1981, p. 251.
[46] J. S. Felton, M. G. Knize, C. Wood, B. J. Wuebbles, S. K. Healy, D. H. Stuermer, L. F. Bjeldanes, B. J. Kimble and F. T. Hatch, *Carcinogenesis,* 1984, **5,** 95.
[47] W. A. Hargraves and M. W. Pariza, *Cancer Res.,* 1983, **43,** 1467.
[48] Z. Yamaizumi, T. Shiomi, H. Kasai, S. Nishimura, Y. Takahashi, M. Nagao and T. Sugimura, *Cancer Lett.,* 1980, **9,** 75.
[49] K. Yamaguchi, K. Shudo, T. Okamoto, T. Sugimura and T. Kosuge, *Gann,* 1980, **71,** 745.
[50] K. Yamaguchi, K. Shudo, T. Okamoto, T. Sugimura and T. Kosuge, *Gann,* 1980, **71,** 743.
[51] T. Matsumoto, D. Yoshida and H. Tomita, *Cancer Lett.,* 1981, **12,** 105.
[52] T. Kawabata, H. Ohshima, J. Uibu, M. Nakamura, M. Matsui and M. Hamano, in *Naturally Occurring Carcinogens-Mutagens and Modulators of Carcinogenesis,* E. C. Miller, J. A. Miller, I. Hirono, T. Sugimura and S. Takayama (eds.), Japan Sci. Soc. Press, Tokyo/Univ. Park Press, Baltimore, 1979, p. 195.
[53] Y. Ohnishi, T. Kinouchi, Y. Manabe, H. Tsutsui, H. Otsuka, H. Tokiwa and T. Otofuji, in M. D. Waters *et al.* (eds.), *Short-Term Genetic Bioassays in the Evaluation of Complex Environmental Mixtures,* Plenum Press, New York, 1983.
[54] C. Y. Wang, A. M. Pamukcu and G. T. Bryan, in *Ecological Perspectives on Carcinogens and Cancer Control,* C. C. Stock, L. Santamaria, P. Mari and S. Gorini (eds.), Karger, Basel, 1976, Vol. 4, p. 565.

[55] L. S. Bjeldanes and G. W. Chang, *Science,* 1977, **197,** 577.
[56] T. Sugimura, M. Nagao, T. Matsushima, T. Yahagi, Y. Seino, A. Shirai, M. Sawamura, S. Natori, K. Yoshihira, M. Fukuoka and M. Kuroyanagi, *Proc. Japan Acad.,* 1977, **53B,** 194.
[57] A. A. Hardigree and J. L. Epler, *Mutation Res.,* 1978, **58,** 231.
[58] J. T. MacGregor and L. Jurd, *Mutation Res.,* 1978, **54,** 297.
[59] J. P. Brown and P. S. Dietrich, *Mutation Res.,* 1979, **66,** 223.
[60] J. P. Brown, *Mutation Res.,* 1980, **75,** 243.
[61] M. Nagao, N. Morita, T. Yahagi, M. Shimizu, M. Kuroyanagi, M. Fukuoka, K. Yoshihira, S. Natori, T. Fujino and T. Sugimura, *Environ. Mutagenesis,* 1981, **3,** 401.
[62] L. E. Sacks and J. T. MacGregor, *Mutation Res.,* 1982, **95,** 191.
[63] C. A. Elliger, P. R. Henika and J. T. MacGregor, *Mutation Res.,* 1984, **135,** 77.
[64] G. Tamura, C. Golc, A. Ferro-Luzzi and B. N. Ames, *Proc. Natl. Acad. Sci. U.S.A.,* 1980, **77,** 4961.
[65] J. P. Brown, *CRC Crit. Rev. Food Sci. Nutr.,* 1977, **8,** 229.
[66] J. H. Carver, A. V. Carrano A. and J. T. MacGregor, *Mutation Res.,* 1983, **113,** 45.
[67] D. E. Amacher, S. Paillet and V. A. Ray, *Mutation Res.,* 1979, **64,** 391.
[68] M. L. Meltz and J. T. MacGregor, *Mutation Res.,* 1981, **88,** 317.
[69] J. C. M. van der Hoeven, I. M. Bruggeman and F. M. H. Debets, *Mutation Res.,* 1984, **136,** 9.
[70] H-U. Aeschbacheı, H. Meier and E. Ruch, *Nutrition and Cancer,* 1982, **4,** 90.
[71] R. K. Sahu, R. Basu and A. Sharma, *Mutation Res.,* 1981, **89,** 69.
[72] J. T. MacGregor, C. M. Wehr, G. D. Manners, L. Jurd, J. L. Minkler and A. V. Carrano, *Mutation Res.,* 1983, **124,** 255.
[73] G. F. A. Cea, K. F. C. Etcheberry and F. N. Dulout, *Mutation Res.,* 1983, **119,** 339.
[74] A. M. Ambrose, D. J. Robbins and F. DeEds, *J. Pharm. Assoc.,* 1952, **41,** 119.
[75] I. Hirono, I. Ueno, S. Hosaka, H. Takanashi, T. Matsushima, T. Sugimura and S. Natori, *Cancer Lett.,* 1981, **13,** 15.
[76] H. Takanashi, S. Aiso, I. Hirono, T. Matsushima and T. Sugimura, *J. Food Safety,* 1983, **5,** 55.
[77] C. Y. Wang, C. W. Chiu, A. M. Pamukcu and G. T. Bryan, *J. Natl. Cancer Inst.,* 1976, **56,** 33.
[78] B. L. Van Duuren and B. M. Goldschmidt, *J. Natl. Cancer Inst.,* 1976, **56,** 1237.
[79] D. Saito, A. Shirai, T. Matsushima, T. Sugimura and I. Hirono, *Teratog. Carcinog. Mutagenesis,* 1980, **1,** 213.
[80] S. Hosaka and I. Hirono, *Gann,* 1981, **72,** 327.
[81] K. Morino, N. Matsukura, T. Kawachi, H. Ohgaki, T. Sugimura and I. Hirono, *Carcinogenesis,* 1982, **3,** 93.
[82] A. M. Pamukcu, S. Yalciner, J. F. Hatcher and G. T. Bryan, *Cancer Res.,* 1980, **40,** 3468.
[83] Y. Takahashi, M. Nagao, T. Fujino, Z. Yamaizumi and T. Sugimura, *Mutation Res.,* 1979, **68,** 117.
[84] Y. Seino, M. Nagao, T. Yahagi, T. Sugimura, T. Yasuda and S. Nishimura, *Mutation Res.,* 1978, **58,** 225.
[85] M. Uyeta, S. Taue and M. Mazaki, *Mutation Res.,* 1981, **88,** 233.
[86] M. Mazaki, T. Ishii and M. Uyeta, *Mutation Res.,* 1982, **101,** 283.
[87] C. S. Yang, *Cancer Res.,* 1980, **40,** 2633.
[88] S-H. Lu, A-M. Camus, L. Tomatis and H. Bartsch, *J. Natl. Cancer Inst.,* 1981, **66,** 33.
[89] M-H. Li, S-H., Lu, C. Ji, Y. Wang, M. Wang, S. Cheng and G. Tian, in *Genetic and Environmental Factors in Experimental and Human Cancer,* H. V. Gelboin, B. MacMahon, T. Matsushima, T. Sugimura, S. Takayama and H. Takebe (eds.), Japan Sci. Soc. Press, Tokyo, 1980, p. 139.
[90] M. Fukuoka, K. Yoshihira, S. Natori, K. Sakamoto, S. Iwahara, S. Hosaka and I. Hirono, *J. Pharm. Dyn.,* 1980, **3,** 236.
[91] J. A. Evans and J. Mason, *Nature,* 1965, **208,** 913.
[92] A. M. Pamukcu and J. M. Prince, *J. Natl. Cancer Inst.,* 1969, **43,** 275.
[93] J. C. M. van der Hoeven, W. J. Lagerweij, M. A. Pothumus, A. van Veldhuizen and H. A. J. Holterman, *Carcinogenesis,* 1983, **4,** 1587.
[94] I. Hirono, K. Yamada, H. Niwa, Y. Shizuri, M. Ojika, S. Hosaka, T. Yamaji, K. Wakamatsu, H. Kigoshi, K. Niiyama and Y. Uosaki, *Cancer Lett.,* 1984, **21,** 239.

[95] I. Hirono, H. Mori, K. Yamada, Y. Hirata, M. Haga, H. Tatematsu and S. Kanie, *J. Natl. Cancer Inst.*, 1977, **58,** 1155.

[96] H. Yamanaka, M. Nagao, T. Sugimura, T. Furuya, A. Shirai and T. Matsushima, *Mutation Res.*, 1979, **68,** 211.

[97] I. Hirono, H. Mori, M. Haga, M. Fujii, K. Yamada, Y. Hirata, H. Takanashi, E. Uchida, S. Hosaka, I. Ueno, T. Matsushima, K. Umezawa and A. Shirai in *Naturally Occurring Carcinogens-Mutagens and Modulators of Carcinogenesis,* E. C. Miller, J. A. Miller, I. Hirono, T. Sugimura and S. Takayama (eds.), Japan Sci. Soc. Press. Tokyo/Univ. Park Press, Baltimore, 1979, p. 79.

[98] I. Hirono, M. Haga, M. Fujii, S. Matsuura, N. Matsubara, M. Nakayama, T. Furuya, M. Hikichi, H. Takanashi, E. Uchida, S. Ohsaka and I. Ueno, *J. Natl. Cancer Inst.*, 1979, **63,** 469.

[99] B. Toth, in *Naturally Occurring Carcinogens-Mutagens and Modulators of Carcinogenesis,* E. C. Miller, J. A. Miller, I. Hirono, T. Sugimura and S. Takayama (eds.), Japan Sci. Soc. Press, Tokyo/Univ. Park Press, Baltimore, 1979, p. 57.

[100] B. Toth, *Proc. Fed. Am. Soc. Exptl. Biol.*, 1980, **39,** 884.

[101] E. G. Rogan, B. A. Walker, R. Gingell, D. L. Nagel and B. Toth, *Mutation Res.*, 1982, **102,** 413.

[102] B. Toth, D. Nagel and A. Ross, *Br. J. Cancer,* 1982, **46,** 417.

[103] S. R. Tannenbaum, in *Naturally Occurring Carcinogens-Mutagens and Modulators of Carcinogenesis,* E. C. Miller, J. A. Miller, I. Hirono, T. Sugimura and S. Takayama (eds.), Japan Sci. Soc. Press, Tokyo/Univ. Park Press, Baltimore, 1979, p. 211.

[104] R. A. Scanlan, *Cancer Res. (Suppl.),* 1983, **43,** 2435s.

[105] K. Wakabayashi, M. Ochiai, H. Saitô, M. Tsuda, Y. Suwa, M. Nagao and T. Sugimura, *Proc. Natl. Acad. Sci. U.S.A.*, 1983, **80,** 2912.

[106] M. Ochiai, K. Wakabayashi, M. Nagao and T. Sugimura, *Gann,* 1984, **75,** 1.

[107] B. G. Piacek-Llanes and S. R. Tannenbaum, *Carcinogenesis,* 1982, **3,** 1379.

[108] D. Yang, S. R. Tannenbaum and G. Büchi, in *Sci. Proc. Am. Assoc. Cancer Research, Toronto,* 9-12 May, 1984, p. 101.

[109] T. Osawa, H. Ishibashi, M. Namiki, M. Yamanaka and K. Namiki, *Mutation Res.*, 1981, **91,** 291.

[110] T. Osawa, M. Namiki and K. Namiki, *Agric. Biol. Chem.*, 1982, **46,** 3105.

[111] H. Marquardt, F. Rufino and J. H. Weisburger, *Science,* 1977, **196,** 1000.

[112] J. H. Weisburger, H. Marquardt, N. Hirota, H. Mori and G. M. Williams, *J. Natl. Cancer Inst.*, 1980, **64,** 163.

[113] K. Wakabayashi, M. Nagao, M. Ochiai, T. Tahira and T. Sugimura, A mutagen precursor in Chinese cabbage, which becomes mutagenic after nitrite treatment, in *Proc. 43rd Annual Meeting, Japan Cancer Association, Fukuoka,* 1984, in the press.

[114] M. Nagao, Y. Takahashi, H. Yamanaka and T. Sugimura, *Mutation Res.*, 1979, **68,** 101.

[115] H. Y. Aeschbacher and H. P. Würzner, *Toxicol. Lett.*, 1980, **5,** 139.

[116] M. Nagao, Y. Suwa, H. Yoshizumi and T. Sugimura, *Mutagens in Coffee,* Banbury Reports, 17, 1984, in the press.

[117] F. Nakasato, M. Nakayasu, Y. Fujita, M. Nagao, M. Terada and T. Sugimura, *Mutation Res.*, 1984, **141,** 109.

[118] L. F. Bjeldanes and H. Chew, *Mutation Res.*, 1979, **67,** 367.

[119] H. Kasai, K. Kumeno, Z. Yamaizumi, S. Nishimura, M. Nagao, Y. Fujita, T. Sugimura, H. Nukaya and T. Kosuge, *Gann,* 1982, **73,** 681.

[120] H. Ristow and G. Obe, *Mutation Res.*, 1978, **58,** 115.

[121] M. Nagao, Y. Takahashi, K. Wakabayashi and T. Sugimura, *Mutation Res.*, 1981, **88,** 147.

[122] C. Loquet, G. Toussaint and J. Y. Letalaer, *Mutation Res.*, 1981, **88,** 155.

[123] J. S. K. Lee and L. Y. Y. Fong, *Food Cosmet. Toxicol.*, 1979, **17,** 575.

6

Mutagenic activity in Dutch river water and its biological significance for fish

C. F. van KREIJL and **W. SLOOFF**, National Institute for Public Health and Environmental Hygiene (RIVM), P.O. Box 150, 2260 AD Leidschendam, The Netherlands

6.1 INTRODUCTION

In The Netherlands, the increasing pollution of river water by xenobiotic compounds has been noted for more than 50 years [1]. Broad public concern, however, stems mainly from the first observation of massive fish kills that could be clearly linked to a specific organic pollutant, i.e. the endosulfan incident in the river Rhine in 1969. Since then, considerable effort has been directed to the qualitative and quantitative investigation of organic compounds in Dutch river waters. Several chemical surveys have been made, revealing the presence of hundreds of organic compounds in the two main rivers, the Rhine and the Meuse. Although most of these compounds were present in concentration below 1 μg/l., it should be stressed that they represent only a minor fraction ($\leqslant$10%) of the total organic material, owing to the selectivity of the analytical methods. Furthermore, several of them were known carcinogens and mutagens. This gave rise to increasing concern about the possible chronic genotoxic effects of polluted river waters on the aquatic organisms inhabiting them, and to the potential health risks associated with their use for drinking water.

In this chapter we will discuss more specifically the presence and detection of organic mutagens and carcinogens in Dutch river water, and their significance for aquatic life. Special attention will be paid to the detection and characterization of unknown mutagens by a combined XAD-concentration/Salmonella mutagenicity assay, to the marked bio-concentration of mutagens in the bile fluid of native fish (bream *Abramis brama*), and to the possible genotoxic effects observed in these fish.

6.2 CARCINOGENS AND MUTAGENS IDENTIFIED IN RHINE WATER

Recently, we have examined over 500 compounds for which CAS registry numbers and reliable references regarding their identification in Rhine water could be obtained [2]. These compounds were screened for possible toxic and genotoxic properties by computerized searches of the individual compounds in the toxicological data bases (RTECS)* of the NIH-EPA Chemical Information System (CIS)†. In addition, for carcinogenic and genotoxic properties, the monogrphas of the WHO International Agency for Research on Cancer (IARC)§ and the data base of the Environmental Mutagen Information Centre (EMIC)‡ were studied.

In Table 6.1, the most important carcinogenic and mutagenic information is summarized. Eighteen compounds in IARC classification Ia and 31 compounds in classification Ib could be detected. In addition the RTECS data base provided information about some 50 suspected carcinogens which are currently being (re)tested in the U.S.A., and another 30 compounds for which there are so far only non-evaluated literature reports of tumorigenic activity (data not shown). Of the mutagens, information was available for about 150 of the compounds found in Rhine water and many of these also fell in groups of carcinogens just mentioned. The genotoxic effects observed concerned biological systems of different complexity (prokaryotes, eukaryotes *in vitro* or *in vivo*) and included different genetic end points (gene mutation, chromosomal abnormalities, DNA damage, etc.). However, for the majority of these genotoxic compounds, the information available was rather limited, and did not fulfil the criteria for a 'qualitative phase mutagen' that were recommended by a committee of the

Table 6.1 – Organic carcinogens and mutagens identified in Rhine water

Compound	CAS-number	Carcinogenicity††	Mutagenicity§§
(Ia) *o*-Anisidine	90-04-0	++	+/–
Benzene	71-43-2	++	+/–
Benzene, hexachloro-	118-74-1	++	+/–
Benzidine, 3,3′-dichloro-*	91-94-1	++	++
Benzo(*a*)pyrene	50-32-8	++	++
Benzo(*b*)fluoranthene	205-99-2	++	+/–
Biphenyl, polychlorinated	1336-36-3	++	+/–
Carbon tetrachloride	56-23-5	++	+/–
Chloroform	67-66-3	++	+/–
DDT	56-29-3	++	+/–
Ethane, 1,2-dibromo-	106-93-4	++	++
Ethane, 1,2-dichloro-	107-06-2	++	++
Indeno(1,2,3-*cd*)pyrene	193-39-5	++	+/–
2-Naphthylamine*	91-59-8	++	++

**Registry of Toxic Effects of Chemical Substances,* NIOSH

†NIH-EPH Chemical Information System

§WHO International Agency for Research on Cancer, Monograph and Annual Reports 1972–84, IARC, Lyon, France.

‡Environmental Mutagen Information Center, Oak Ridge National Laboratory, P.O. Box 7, Oak Ridge, TN, U.S.A.

Table 6.1 – *continued*

Compound	CAS-number	Carcinogenicity†	Mutagenicity §
Phenacetin	62-44-2	++	+
Phenol, 2,4,6-trichloro-	88-06-2	++	+/–
Phthalic acid, bis(2-ethylhexyl)ester*	117-81-7	++	+/–
o-Toluidine	95-53-4	++	+
(Ib) Adipic acid, bis(2-ethylhexyl)ester*	103-23-1	+	+/–
Aldrin	309-00-2	+	+/–
Aniline	62-53-3	+	+/–
o-Anisidine, 5-nitro-	99-59-2	+	+/–
Azobenzene	103-33-3	+	+/–
Butadiene, hexachloro-	87-68-3	+	+/–
Chrysene	218-01-9	+	+/–
Cyclohexane, hexachloro- (α)	319-84-6	+	+/–
Cyclohexane, hexachloro- (γ)	58-89-9	+	+/–
DDD (*p,p'*)	72-54-8	+	+/–
DDE (*p,p'*)	72-55-9	+	+/–
Dieldrin	60-57-1	+	+/–
Ethane, hexachloro-	67-72-1	+	+/–
Ethane, pentachloro-	76-01-7	+	+/–
Ethane, 1,1,2,2-tetrachloro-	79-34-5	+	+/–
Ethane, 1,1,2-trichloro-	79-00-5	+	+/–
Ethene, 1,1,-dichloro-	75-35-4	+	++
Ethene, tetrachloro-	127-18-4	+	+/–
Ethene, 1,1,2,-trichloro-	79-01-6	+	+/–
Heptachlor	76-44-8	+	+/–
Heptachlor epoxide	1024-57-3	+	+/–
Phenol, 2-phenyl-*	90-43-7	+	+/–
Styrene	100-42-5	+	++
Styrene oxide	96-09-3	+	++
Toluene, 2,4-dinitro-	121-14-2	+	++
Toluene, α-chloro-	100-44-7	+	++
Toluene, α,α-dichloro-	98-87-3	+	+/–
Toluene, α,α,α-trichloro-	98-07-7	+	+/–
o-Toluenesulphonamide*	88-19-7	+	+/–
o-Toluidine, 5-chloro-*	95-79-4	+	+/–
o-Toluidine, 5-nitro-*	99-55-8	+	+/–
(II) 1-Naphthylamine*	134-32-7	+/–	++
Naphthalene, 2-nitro-*	581-89-5	+/–	++
Propene, 1,1,2,3-tetrachloro-*	10436-39-2		+
Quinoline	91-22-5		++
o-Toluidine, 4-chloro-*	95-69-2	+/–	+

*Compound not identified as such, but as isomeric mixture.

††++ = Classified by IARC as carcinogen (animal or human) with sufficient evidence.

+ = Classified as animal carcinogen with limited evidence; evaluated (IARC, NCI) positive results not (yet) fulfilling criteria for sufficient evidence with regard to multiple species or different exposure routes.

+/– = Experimental data classified by IARC as inadequate for evaluation of carcinogenicity.

§§++ = Compounds for which there is sufficient evidence to regard them as both positive in short-term tests according to the IARC criteria, and mutagenic according to the recommendations of the Netherlands Health Council.

+ = Compounds fulfilling either one of the criteria above.

+/– = Compounds for which insufficient evidence exists regarding the criteria above; results requiring further evaluation or testing.

Netherlands Health Council [3], or those for 'sufficient evidence in short-term tests' as defined by IARC. This is illustrated in Table 6.1, where of the 49 recognized carcinogens in groups Ia and Ib only 12 can be considered as 'sufficiently' mutagenic. Similarly only 5 mutagens fulfilling the criteria above were found among the remaining 100 'genotoxic' compounds (Table 6.1, group II).

Table 6.2 – Salmonella mutagens identified in Rhine water

Compound	CAS-number	Dose-level†	Strain§	S9‡
PAHs				
Benzo(*a*)pyrene	50-32-8	+++	100/98	+
Benzo(*b*)fluoranthene	205-99-2	+++	100/98	+
Benzo(*k*)fluoranthene	207-08-9	+++	100/98	+
Indeno(1,2,3-*cd*)pyrene	193-39-5	+++	100/98	+
Aromatic amines, azo-compounds				
Azobenzene	103-33-3	++	100	+
Aniline, 3,4-dimethyl-*	95-64-7	+++	100	+
Aniline, dimethyl-(2,4/2,5)*		+	100	+
Aniline, *p*-chloro-	106-47-8	+	100/98	+
o-Anisidine	90-04-0	+	98	+
Benzidine, 3,3′-dichloro-*	91-94-1	++	100/98	+
2-Naphthylamine*	91-59-8	+++	98/100	+
1-Naphthylamine*	134-32-7	++	98/100	+
o-Toluidine, 4-chloro-*	95-69-2	+++	100	+
o-Toluidine, 5-nitro-*	99-55-8	+	100/98	–
Nitro-aromatics, N-*heterocyclics*				
Benzene, *m*-dinitro-*	99-65-0	++	100/98	–
Naphthalene, 1-nitro-*	86-57-7	+++	100	–
Naphthalene, 2-nitro-*	581-89-5	+++	100	–
Toluene, dinitro-(2,5/3,5)*	(2x)	++	100/98	–
Toluene, dinitro-(2,3/2,4/2,6/3,4)	(4x)	+	100	–
Acridine	260-94-6	++	1537/100	+
Benzo(*f*)quinoline	230-27-3	++	1537/100	+
Quinoline	91-22-5	++	1537/100	+
Quinoline, methyl-(4/6/7)	(3x)	++	.1537/100	+
Halogenated compounds				
Ethane, 1,2-dibromo-	106-93-4	++	100	–
Ethane, 1,2-dichloro-	107-06-2	+	100	+
Propane, 1,2,3-trichloro-*	96-18-4	++	100	–
2-Propanol, 1,3-dichloro-	96-23-1	+	100	–
Propene, 2,3-dichloro-*	78-88-6	++	100	–
Propene, 1,3-dichloro-*	542-75-6	+	100	–
Propene, 1,1,2,3-tetrachloro-*	10436-39-2	++	100	–
Toluene, α,α-dichloro-	98-87-3	++	100	+
Toluene, α,α,α-trichloro-	98-07-7	++	100/98	+
Miscellaneous				
Anthraquinone	84-65-1	+	100	–
Styrene	100-42-5	+	100	+
Styrene oxide	96-09-3	+	100	–

*Compound not identified as such, but as isomeric mixture.
†Lowest positive dose observed; + 100–1000; ++ 10–100; +++ < 10 (μg/plate).
§Bacterial strain(s) with which lowest positive dose-level is observed.
‡Requirement for metabolic activation at lowest positive dose-level.

Finally, we have also examined which of the 150 'genotoxic' compounds were reported to be mutagenic in bacterial test systems such as the Salmonella mutagenicity assay [4], and at what dose levels. Of the 90 or so reported Salmonella mutagens, less than half were active at levels below 1 mg/plate. In Table 6.2 these compounds are grouped according to their overall chemical class. It is shown that the most active Salmonella mutagens (active at $< 10\ \mu g$/plate) are, besides the well-known polycyclic aromatic hydrocarbons (PAHs), aromatic amines and nitro-compounds. The implications of this finding for the mutagenic activities observed in organic concentrates of Rhine water will be discussed later.

6.3 MUTAGENICITY OF DUTCH RIVER WATERS

The organic compounds identified in surface water represent only a minor fraction of the total organic material present, and this situation is not likely to change dramatically within the next few years. This has prompted many investigators to examine the remaining fraction of as yet unidentified organic compounds by using direct biological assays. In this respect special attention has been paid to the presence of mutagenic substances.

In the Netherlands, the first observations on the mutagenicity of river water were made by Prein *et al.* [5], who screened laboratory fish (mudminnow, *Umbra pygmaea*) for the presence of chromosome breaks in gill cells after 11 days exposure to Rhine water. Alink *et al.* [6] confirmed and extended these cytogenetic results by showing up to 3-fold increases in sister-chromatid exchanges in both testicular and gill cells of mudminnows exposed to Rhine water. Similar observations were reported by Hooftman and Vink [7], and later by Van der Gaag *et al.* [8], who used the tropical fish *Nothobranchius rachowi.*

However, the cytogenetic effects in fish were usually rather small and often difficult to reproduce [8]. Therefore, the use of methods for concentration of organic compounds in water, combined with short-term bacterial mutagenicity tests, has received increasing attention over the past five years. As argued by Yamasaki and Ames [9], adsorption on XAD-resins might be very specific for the isolation of mutagenic substances from aqueous solutions. This concentration method, in combination with the Salmonella mutagenicity test [4], has been adopted worldwide for the detection of unknown mutagens in all kinds of waters. Mutagenicity results thus accumulated in The Netherlands will be discussed below.

6.3.1 Geographical differences in the observed mutagenicity

Figure 6.1a shows the locations where samples from Dutch surface waters have been taken by different investigators during the last six years. They include the river Rhine and its branches, the river Meuse, some lakes and the North Sea coastal area. To allow discrimination between Dutch and foreign sources of pollution in the two major rivers, attention was also paid to some locations in Germany and Belgium. At first, positive mutagenic effects were reported with strain TA100 by testing petrol-ether extracts of 6 litres of Rhine water (location no. 5) in the Salmonella mutagenicity assay [5].

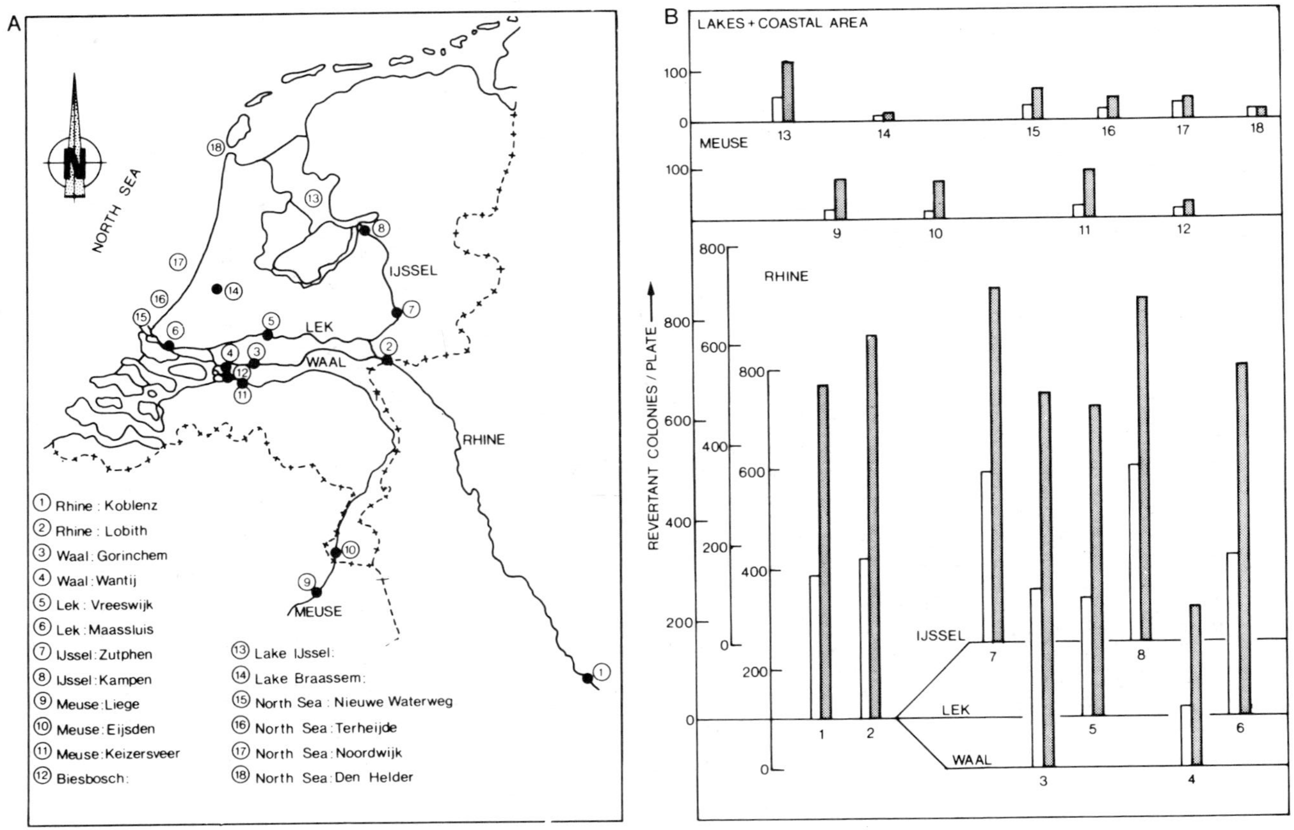

Fig. 6.1 – Mutagenic activity in Dutch surface waters. A, Locations of sampling. B, Mutagenic activity corresponding to concentrates of 1 litre of water, as measured with strain TA98. Open and closed bars represent the absence and presence of rat liver S9 respectively. Average values are shown, after subtraction of background revertants. (Modified after Van Kreijl *et al.* [10], Kool *et al.* [11], Van Hoof and Verheijden [15], Poels [13], Van der Gaag *et al.* [14], Slooff and Van Kreijl [12], Van Kreijl, unpublished work.)

Two years later, Van Kreijl *et al.* [10] and Kool *et al.* [11] demonstrated not only the effectiveness of the combined XAD/Salmonella method for Dutch surface waters, but also that significant mutagenic activity could be measured in concentrates of as little as 50 ml of Rhine water (nos. 2–8). This activity, moreover, was observed primarily with the tester strain TA98, and was found to increase strongly on addition of a metabolic activation system (rat liver S9).

In recent years, these results have been confirmed and extended by investigators from our institute, as well as from the Testing and Research Institute of the Netherlands Waterworks. Using strain TA98, 250-ml XAD-concentrates and repeated sampling, Slooff and Van Kreijl [12] showed that nearly all the Rhine samples (nos. 2, 3, 5) contained both direct and indirect frameshift mutagens (+S9 > –S9). In contrast, no significant mutagenic activity was detected in the lake sample (no. 14) and in most of the river Meuse samples (nos. 10, 11). Using a somewhat modified XAD-method and concentrates of up to 3 litres of water, Poels [13] and Van der Gaag *et al.* [14] demonstrated the presence of Salmonella TA98 mutagens in Rhine water (nos. 2, 5), Meuse water (no. 11) and Lake IJssel water (no. 13). In addition, positive results with the TA100 tester strain were reported, but the activity was much lower than with TA98. Other results regarding the mutagenicity of Meuse water in both Belgium and The Netherlands were reported by Van Hoof and Verheyden [15], using dichloromethane extracts, and in a recent RIWA report [16].

In Fig. 6.1b a schematic survey is presented of the average mutagenic activities (TA98) corresponding to 1 litre of water, as observed by various investigators (published and unpublished data). Consistent differences were found between the different surface waters. When the waters are ranked in order of decreasing mutagenic potency, the following sequence is obtained: river Rhine and branches ≫ Lake IJssel > river Meuse > North Sea > Lake Braassem. Besides these quantitative aspects, the following general conclusions can be drawn.

(1) Mutagenic activity can be observed in XAD concentrates of most surface waters in The Netherlands. The Salmonella mutagenicity is observed primarily with the bacterial strain TA98, and increases upon addition of rat liver S9. The latter effect is most pronounced with Meuse water.
(2) Additional but minor TA98 activity can be observed in river Rhine, Meuse, and Lake IJssel concentrates when the XAD-adsorption is done under acidic conditions (pH 2–3).
(3) Differences in mutagenic activity may occur in different regions or in different areas of the same surface water. Local circumstances, for instance the operation of weirs, may result in an increased self-purification process of the river at the time of sampling, as observed for the river Lek [12, 13]. Dilution processes also play an important role in diminishing the mutagenic activity, as is shown for Lake IJssel and the North Sea coast.
(4) The mutagenic activity observed in the rivers Rhine and Meuse does not usually change very much after they enter The Netherlands. This suggests that most of the mutagenic pollution is of foreign origin. The major sources of mutagenic pollution in the river Meuse seem to be located downstream

of Namen and Liège in Belgium [15, 16]. The preliminary results of a study of the river Rhine under the auspices of the International Rhine Committee suggest a location upstream of Koblenz as the main source of mutagenic pollution.

6.3.2 Temporal difference in the observed mutagenicity

So far, only a few river Rhine locations have been studied frequently enough to allow more accurate estimates of the variation in mutagenic activity with time. For locations 2, 3 and 5 the mutagenicity of comparable organic concentrates (XAD/TA98) was found to vary from 5- to 10-fold during a period of one year [12–14]. Although these variations in activity are largely due to seasonal effects such as temperature and river flow, other factors such as changes in chemical discharges or use of weirs may also contribute. Variations due to inhomogeneous spreading of mutagens over the water stream, or to sampling and testing procedures are thought to be of minor importance [12].

6.3.3 Specificity of the mutagenic activity in Rhine water

When XAD concentrates of Rhine water (nos. 2, 3) were tested with all 5 bacterial tester strains, as originally recommended by Ames *et al.* [4], activity was primarily observed with the strains TA1538 and TA98 (see Fig. 6.2). This indicates the presence of both direct and indirect frameshift mutagens, and most of these seem not to require the plasmid pKM101 for mutagenic effect. Recently, Maron and Ames [17] have recommended that strains TA1535, TA1537 and TA1538 be replaced by the newly developed plasmid-containing strains TA97 and TA102.

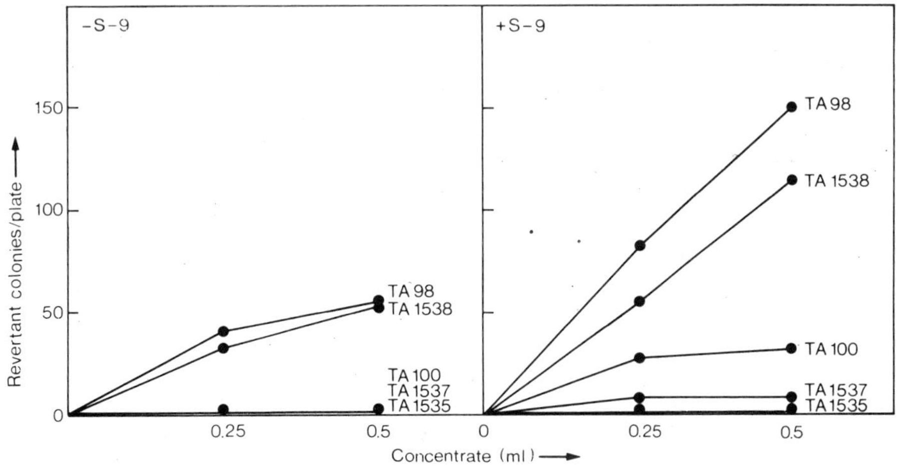

Fig. 6.2 – Mutagenic activity of Rhine water concentrates with different Salmonella tester strains. XAD-concentrates corresponding to 250 ml and 500 ml of Rhine water (no. 3) were tested in the presence and absence of rat liver S9. Each point represents the average value of 3 plates, corrected for the spontaneous revertants (68–70 for TA100 and 5–20 for the other strains). (Modified after Van Kreijl *et al.* [18, 19].)

Unlike the other strains, TA102 carries its histidine-mutation in an AT-stretch of DNA, located on a multicopy plasmid, which renders it particularly sensitive to oxidative genotoxic agents. However, no significant mutagenic activity was detected in the Rhine water concentrates by using these two new strains (Van Kreijl, unpublished work).

Relatively little information is available with regard to mutagenic effects in eukaryotic systems, owing to the toxicity of the water samples. Nevertheless, as shown in Table 6.3, positive results have occasionally been obtained when water samples were tested for cytogenetic effects in CHO-cells *in vitro* [18, 19]. In contrast to Meuse water, Rhine water samples induced significant increases in the number of chromosomal aberrations as well as sister-chromatid exchanges (SCEs). Chromosomal aberrations (chromatid type) were observed only in the absence of rat liver S9, whereas induction of SCEs was observed mainly in the presence of S9. Without the metabolic activation system, the SCE analysis was

Table 6.3 – Induction of chromosome abberrations and sister-chromatid exchanges in CHO-cells in vitro by river water concentrates

For chromosomal aberrations, 100 μl of XAD-concentrate corresponds to 500 ml of Rhine water (expt. 1) or 800 ml of Rhine and Meuse water (expt.2). Sampling was at locations 3 and 6. For SCE-analysis 100 μl of sample corresponds to 6 and 5 litres of Rhine water (no. 3) in expts. 3 and 4 respectively. Figures represent the average values ± S.D. of 25 metaphases. Dimethylnitrosamine (DMN) was included as positive control. Mutagenicity results refer to revertants induced with TA98.

Sample			Salmonella mutagenicity		Chromosome aberrations per 50 cells		SCE/metaphase	
			–S9	+S9	–S9	+S9	–S9	+S9
Expt. 1:	none		15	20	3	5	–	–
	Rhine,	150 μl	–	–	12	6	–	–
	Rhine,	300 μl	633	933	6	2	–	–
	Rhine,	500 μl	–	–	14	6	–	–
Expt. 2:	acetone,	100 μl	24	30	1	3	–	–
	Meuse,	25 μl	–	–	1	7	–	–
	Meuse,	50 μl	–	–	2	2	–	–
	Meuse,	100 μl	40	72	7	6	–	–
	Rhine,	25 μl	–	–	4	5	–	–
	Rhine,	50 μl	–	–	10	9	–	–
	Rhine,	100 μl	308	683	19	2	–	–
Expt. 3:	acetone,	100 μl	30	22	–	–	8±2	10±3
	Rhine,	50 μl	337	764	–	–	–	–
	Rhine,	100 μl	–	–	–	–	*	17±5
	DMN (0.04 *M*)		–	–	–	–	–	33±8
Expt. 4:	acetone,	100 μl	30	27	–	–	8±3	8±3
	Rhine,	25 μl	146	383	–	–	12±5	11±3
	Rhine,	50 μl	216	655	–	–	*	14±5
	Rhine,	100 μl	–	–	–	–	*	*
	DMN (0.04 *M*)		–	–	–	–	–	63±9

– not tested; * sample is toxic for the cells.

clearly hampered by the toxicity of the samples. Finally, the testing of Rhine water samples with the fruit fly *Drosophila melanogaster* was found unsuccessful because of the toxicity of the samples (Blijleven and Van der Gaag, personal communication).

6.3.4 Nature of the organic mutagens in Rhine water

The continuous presence of mutagenic activity in organic concentrates of river Rhine water raises some questions with regard to the nature and identity of the mutagens involved. As stated before (Section 6.2), bacterial mutagenicity data are available for about 90 of the organic compounds identified so far, some of which are active at low doses (Table 6.2). Although, theoretically, some of these compounds (or combinations of them) could be responsible for the observed mutagenicity in Rhine water concentrates, several lines of evidence argue against this possibility.

First, the mutagenic samples contain less than 100 μg of organic matter [12] and the mutagens involved do not comprise more than 10% of this, suggesting that the observed mutagenicity must be due to a number of relatively potent mutagens rather than to a large group of very weak mutagens. Secondly, the bacterial strain specificity of the more active mutagens listed in Table 6.2 (active at $<$100 μg/plate) is not in agreement with the mainly frameshift type of activity (TA98, TA1538) observed with Rhine water concentrates. The PAHs and aromatic amines are known to induce frameshift mutations, but many of these compounds are equally or even more active on strain TA100.

Further, nitro-aromatic compounds with $\leqslant$2 aromatic rings have been reported to be preferentially active on strain TA100 [20, 21], and this also holds for the remaining halogenated and miscellaneous compounds. DNA intercalators such as acridine or its derivatives are frameshift mutagens, but they can be detected with the bacterial strain TA1537. Thirdly, the experiment described in Table 6.4 also argues against an important role of the PAHs. Pentane extraction has been described as a specific method for removing PAHs from aqueous dimethylsulphoxide solutions [22]. When this is applied to a Rhine water concentrate, the majority of the mutagenic activity remains in the aqueous phase, while over 90% of added 3H-benzo(*a*)pyrene is recovered in the pentane fraction.

Thus, none of the organic mutagens identified so far can account for major parts of the observed mutagenic activity in organic concentrates of Rhine water. For quantitative comparisons between known and unknown mutagens it is important that the mutagens identified did not escape detection in the mutagenicity test because of low recovery in the concentration procedure. This was verified in several control experiments, including the testing of lyophilized water, the omission of filtration before the XAD adsorption, and elution of the XAD-column with different solvents [11, 18, Van Kreijl, unpublished work]. Only for the more volatile halogenated lower alkanes and alkenes could low recovery not be excluded, but their concentrations in the water and their mutagenic activity render it unlikely that they would have been detected even after full recovery. Therefore considerable amounts of (rather potent) unknown

Table 6.4 – Effect of pentane extraction on the mutagenic activity of a Rhine water concentrate

A Rhine water XAD-concentrate (no. 3) was diluted with distilled water to obtain a DMSO-water ratio of 1:1. A duplicate sample was spiked with ^{3}H-labelled benzo(*a*)pyrene, and both samples were extracted three times with pentane. The spiked sample was assayed for ^{3}H-radioactivity and the unlabelled sample for mutagenic activity. Mutagenicity results correspond to revertants induced with strain TA 98 with or without rat liver S9, and average values for three plates are shown.

Sample	^{3}H-B(*a*)P		Mutagenicity	
	cpm	% of total	–S9	+S9
Control	16	–	42	39
Rhine concentrate before extraction				
0.1 ml	–	–	69	112
0.2 ml	2730	100	93	208
0.4 ml	–	–	139	416
Rhine concentrate after extraction				
0.1 ml	–	–	69	89
0.2 ml	55	2	88	160
0.4 ml	–	–	128	324
Pentane fraction				
0.2 ml	2530	92	–	–

– not tested

mutagens must be present in Rhine water and their combined activity must exceed by far that of the ones identified.

Because chemical analysis still leaves these potent mutagens unidentified, we attempted to characterize these unknown compounds further. Simple evaporation or heating experiments with the concentrates showed that the mutagens are rather stable, non-volatile and slightly polar; gel filtration on Sephadex LH20 according to the method of Concin *et al.* [23] indicated molecular weights ranging from about 100 to 300 [18, 19]. More recently, a first indication of the nature of these compounds was obtained by testing the organic Rhine water concentrates with specific enzyme-deficient strains of Salmonella. As shown in Fig. 6.3, the direct mutagenic activity with strain TA98 is almost completely abolished when the nitroreductase-deficient strains TA98NR and TA98/1,8-DNP$_6$

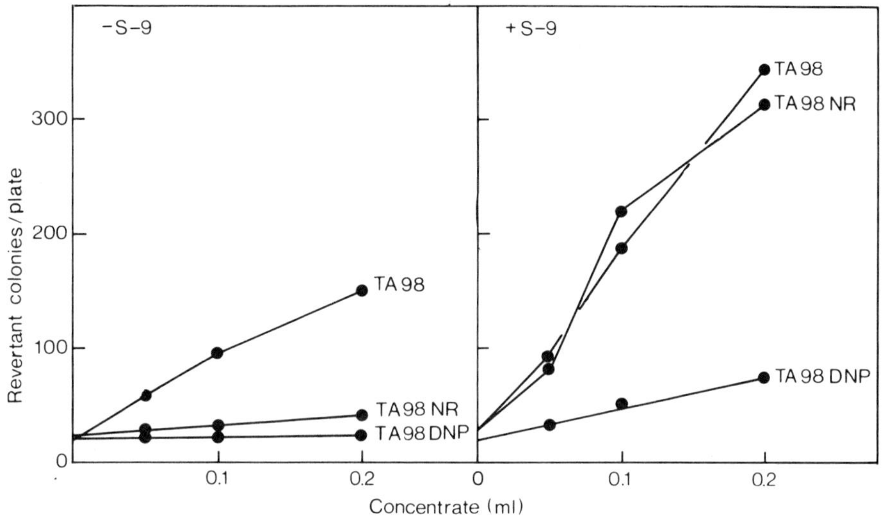

Fig. 6.3 – Mutagenic activity of Rhine water concentrates with nitroreductase-deficient bacterial strains. XAD-concentrates of Rhine water (no. 2) were tested plus or minus rat liver S9. Each point represents the average value of 3 plates.

are used. Given the previous results (Fig. 6.2) and the fact that control experiments demonstrated the capability (~ 70%) of the two enzyme-deficient strains to detect non-nitroreductase-dependent mutagens, this is suggestive of the presence of nitro-aromatics with ⩽ 3 aromatic rings [20].

For the indirect mutagens, a strong reduction of the activity was only observed with strain TA98/1,8-DNP_6, and the activity with strain TA98NR seemed unaffected. This is also considered indicative of the presence of aromatic amines, as will be discussed in Section 6.6.

6.4 MUTAGENIC ACTIVITY IN BILE FLUID OF RHINE FISH

So far, little attention has been paid to the biological significance of the carcinogens and mutagens present in Dutch river waters. To shed some light on the possible impact of the unknown mutagens in this respect, the *in situ* exposure of native fish to these compounds has been investigated by examining the bile fluid. Biliary excretion is known to be an important route for xenobiotics and their metabolites in aquatic organisms [24, 25]. Bio-concentrations of xenobiotics in the gall bladder of fish has been reported for many individual compounds, and bile:water concentration ratios as high as 10^4 have been observed. In fact, this has led Statham *et al.* [26] to propose the monitoring of bile from wild or caged fish to indicate chemical pollution by certain xenobiotics.

It has been demonstrated recently that considerable mutagenic activity can be detected in the bile of native fish – bream, *Abramis brama* – from the river Rhine [27]. As in river water, organic mutagens can be detected in the otherwise toxic bile fluid by dilution with distilled water and subsequent adsorption on XAD-resins. The mutagenic activity of the fish bile in the Salmonella mutagenicity test resembled that of the river water with regard to strain specificity (TA1538 and TA98) and the effect of S9. However, the activity in the bile fluid was at least 10^4 times that of the river water, indicating a strong bio-concentration in the gall bladder. Further, seasonal changes in the level of Rhine water mutagenicity were found to roughly coincide with similar changes in the activity of the fish bile. Addition of β-glucuronidase before the XAD-adsorption step was observed to increase the activity, indicating the presence of additional glucuronide-conjugated mutagens in the bile fluid, which are not recovered in standard XAD extracts. Finally, when the bile fluid of river Meuse fish was examined, no detectable mutagenicity was found, indicating that the ratio of Rhine:Meuse mutagenicity is at least as high for the bile fluids as for the corresponding waters.

These field experiments have now been confirmed in a laboratory study (Van Kreijl *et al.*, in preparation). Young rainbow trout, *Salmo gairdneri*, were placed in large basins and exposed to a continuous flow of Rhine water (Fig. 6.1, no. 2) under controlled conditions. As a result, the original non-mutagenic bile fluid became rapidly mutagenic, reaching maximum potency in about 2 days. Transfer of the fish to clean water after 14 days exposure resulted in an immediate decrease of the bile mutagenicity. When the mutagenic bile extracts were tested with the nitroreductase-deficient strains TA98NR and TA98/1,8-DNP_6, similar results were obtained as with the Rhine water concentrates (see Fig. 6.4), giving further evidence for the presence of nitro-aromatic compounds and aromatic amines.

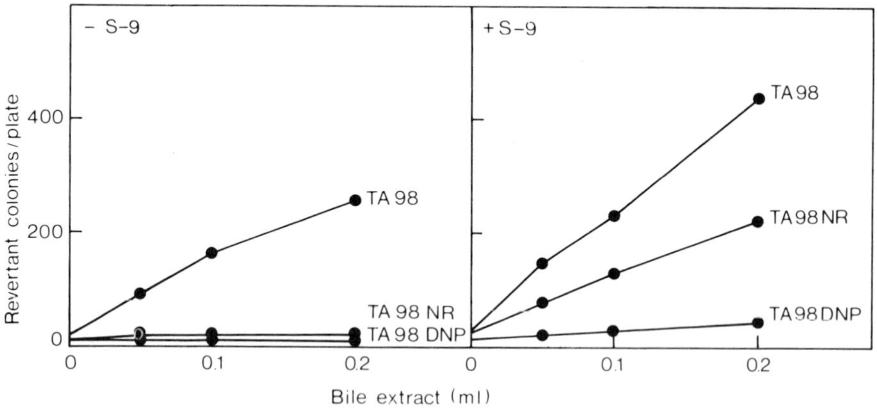

Fig. 6.4 – Mutagenic activity of fish-bile extracts with nitroreductase-deficient bacterial strains. XAD bile extracts were obtained from rainbow trout exposed to Rhine water (no. 2) and 0.1 ml of extract corresponds to 0.02 ml of original bile fluid. Mutagenicity testing was performed with or without rat liver S9. Each point represents the average value of 3 plates.

6.5 BIOLOGICAL EFFECTS ON FISH

Although little is yet known about the possible adverse effects of the mutagens and carcinogens in Dutch river waters on aquatic life, several observations on fish exposed to mutagen-containing river waters may arouse concern. As mentioned in the preceding section, considerable mutagenic activity can be detected in the bile of fish from the river Rhine, indicating that mutagens do enter the fish body. In addition, chromosome breaks and chromatid exchanges have been observed in fish experimentally exposed to Rhine water, as discussed in Section 6.3. To elucidate the possible impact of the mostly unknown mutagens at the individual and population level, attention was therefore paid to the state of health of fish naturally or experimentally exposed to Rhine water.

Under controlled laboratory conditions, both embryo-larval tests [28] and long-term tests (3-18 months) with young rainbow trout (*S. gairdneri*) have been performed on Rhine water after sedimentation of particulate matter [29, 30]. From these laboratory studies it was concluded that Rhine water causes a higher embryo-larval mortality, induces pugheadedness and enlargement of kidney and liver, and results in growth retardation and a lower hematocrit or haemoglobin content. No tumours were observed in 30 fish examined after 18 months exposure. In another experiment 3-4 week old guppies (*Poecilia reticulata*) were exposed for a 6-month period to different doses of organic (XAD) concentrates of Rhine water, with 50 fish per dosage group [31]. No effects were found at a dose corresponding to 10 times the normal Rhine water concentration. When the dose was increased to 33 × normal, impaired swimming performance, hypertrophy of the liver and partial mortality were observed, whereas at doses ⩾ 100 × normal Rhine water concentration, mortality occurred rapidly. Tumours were not found in this group either, in histopathological examination of the total fish.

Table 6.5 – Abnormalities observed in native fish, possibly related to pollution with chemical mutagens and carcinogens
(Modified after Slooff [34, 35])

Location of sampling	Number of fish examined	Prevalence of tumors (%)	Prevalence of deformed fins (%)	Prevalence of pugheadedness (%)
Lake Braassem (14; control)	1404	0.0	4.6	0.3
River Rhine (1,2,3,5,)	4184	1.7	19.3	2.6
Lake IJssel (13)	714	0.0	11.6	1.4
River Meuse (10)	1435	0.7	6.7	0.5

In field studies on the state of health of bream (*A. brama*) collected from various surface waters in The Netherlands (Fig. 6.1, nos. 1, 2, 3, 10, 13, 14) it was found that fish originating from the river Rhine and its tributaries showed higher relative liver weights. The hypertrophic livers were found to contain elevated levels of microsomal enzymes [32], and to have an increased ability to activate premutagens present in Rhine water concentrates [12]. Rhine fish also showed a higher annual mortality and higher fecundity than fish from less polluted areas [33]. Table 6.5 shows a low, but higher, prevalence of tumours

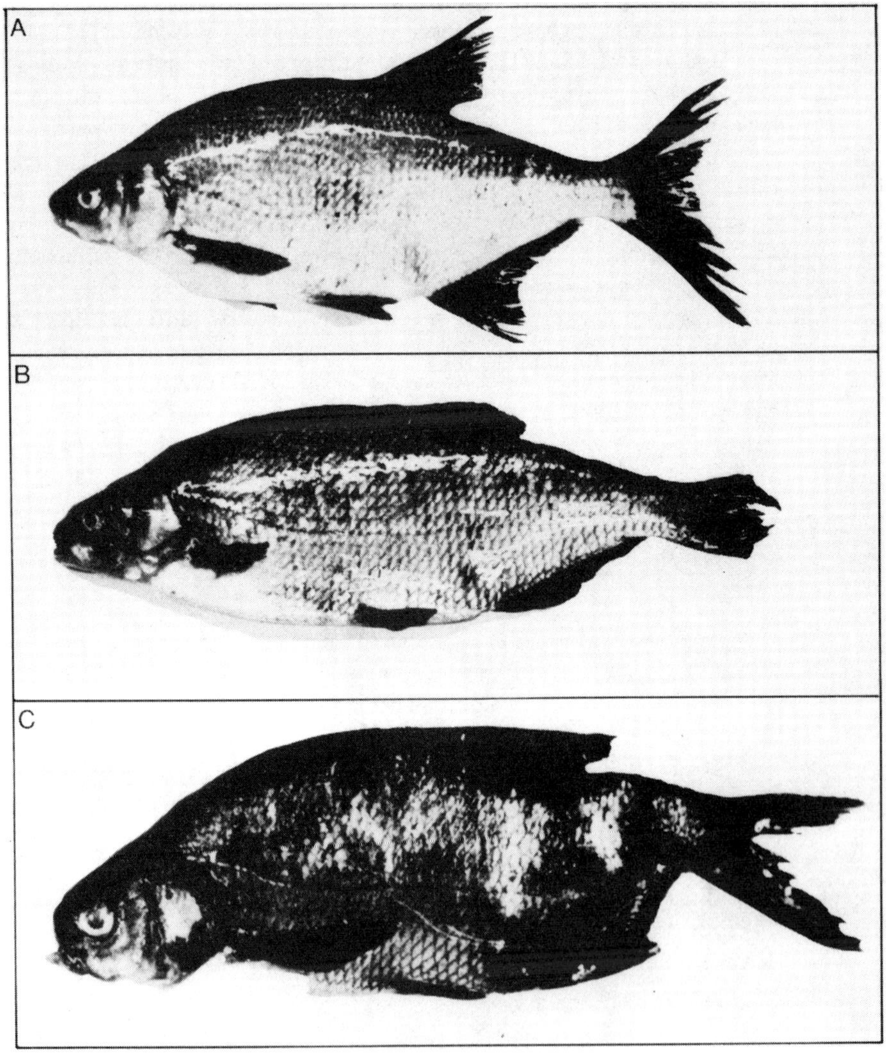

Fig. 6.5 – Skeletal anomalies observed in bream (*Abramis brama*). A, Normal bream. B, Bream with deformations of all fins. C, Pugheaded bream with curvatures in both sagittal and horizontal planes. (After Slooff [34].)

(mostly hepatocellular carcinomas) and a higher frequency of skeletal anomalies such as deformed fins and pugheadedness in Rhine fish, relative to fish from other areas [34, 35]. Examples of these skeletal anomalies are shown in Fig. 6.5.

From the results above, it can be concluded that many deviations from biological and biochemical reference values have been observed in fish exposed to river Rhine water or organic concentrates thereof. A reasonable agreement between the laboratory studies and the field observations is also noted. However, some of the effects are rather non-specific and others are not statistically significant. Therefore, as will be discussed below, no firm conclusions can be drawn at present regarding the reponsibility of the waterborne carcinogens and mutagens for the observed effects.

6.6 DISCUSSION AND CONCLUSIONS

It is clear that considerable amounts of organic mutagens and carcinogens occur in several Dutch surface waters. Some of these have been discovered by chemical analysis, as shown for the river Rhine (Tables 6.1 and 6.2), others indirectly by observing Salmonella mutagenicity in organic concentrates of the water (Fig. 6.1). The latter group demonstrated the usefulness of the combined XAD/mutagenicity-method for monitoring these waters, and revealed a large difference between the mutagenicity of Rhine and Meuse water, which otherwise would not have been noted. It also led to the conclusion that the contribution of the organic mutagens identified so far is of minor importance with respect to the total mutagenicity observed in river Rhine water. The fact that unknown and rather active Salmonella mutagens must be present in at least some of the Dutch surface waters gives rise to three major questions.

(1) What is the identity and source of the compounds responsible for the observed mutagenic activity?
(2) What is the biological significance of the known and unknown mutagens and carcinogens for aquatic life?
(3) Is there a possible health risk involved in using these polluted surface waters as the raw water source for drinking water preparation?

The last question clearly lies beyond the scope of this chapter and has been dealt with elsewhere [36]. Attention will therefore be focused on the first two points.

PAHs have already been suggested to be responsible for the mutagenic activity of both the Rhine and the Meuse [10, 15]. Although the evidence is still indirect, it now appears that nitro-substituted PAHs (nitroarenes) and aromatic amines are the most likely causes of the Rhine water mutagenicity.

In this respect it is interesting to note that, although none of the chemically identified mutagens (either alone or in combination) can account for the observed mutagenicity, the most active and relevant compounds (Tables 6.2 and 6.4) also belong to the same chemical classes. Nitroarenes have been reported to be direct-acting mutagens in Salmonella, but their mutagenicity greatly depends on the active reduction of the nitro-group to hydroxylamine by the bacteria. This explains the strongly reduced mutagenicity with the nitroreductase-deficient strains in the absence of S9 (Fig. 6.3). The preference for TA98 and TA1538 (Fig. 6.2) further indicates that we are probably dealing with nitroarenes containing three or more fused aromatic rings, and not with monocyclic or dicyclic nitro-aromatic compounds [20, 21]. For the indirect mutagenic activity (+S9) the situation is somewhat more complex. A minor part of this activity may be attributed to the nitroarenes (+S9 < −S9), detectable also with TA98NR because the S9 enzymes compensate for the nitroreductase-deficiency. The strongly reduced activity with strain TA98/1,8-DNP_6, however, is suggestive of the presence of aromatic amines. This strain is believed to be blocked in a specific hydroxylamine esterification enzyme [20], and *N*-hydroxylation plus subsequent esterification is known to be an important pathway in the necessary metabolic activation of aromatic amines.

Introduction of organic mutagens in the river Meuse in Belgium has been reported to be downstream of Liège and two coke-oven effluents have been mentioned in this regard [15]. For the river Rhine, a major part of the mutagenicity originates from a location in Germany somewhere upstream of Koblenz, as indicated by the recent study instigated by the International Rhine Committee. It is to be expected, therefore, that a more detailed investigation of the areas concerned, and focusing of analytical chemical efforts on nitroarenes and/or aromatic amines, will rapidly reveal the emission-points and identity of the mutagens involved.

No conclusive evidence exists as to the biological significance of the mutagens and carcinogens in Rhine water. The observed liver enlargements strongly indicate exposure to xenobiotic compounds, but these do not necessarily have to be mutagenic or carcinogenic. Laboratory studies with fish have indicated that Rhine water may induce significant changes at the chromosomal level (Section 6.3); this is supported by both laboratory and field studies which demonstrated a strong bio-concentration of Salmonella mutagens in fish bile (Section 6.4). Since these mutagens seemed similar in nature to those in the river water (Fig. 6.4), it may be concluded that *in situ* exposure of the fish body to increased concentrations of the unknown mutagens indeed takes place.

It can be argued that mutagenic agents may be responsible for the observed skeletal deformations, or more importantly, for the increased embryonal mortality. Many chromosomal aberrations can lead to early zygote death [37], and chronic exposure to genotoxic agents such as ionizing radiation results in increased embryonal mortality, increased (skeletal) abnormalities in young fish, and increased brood size [38]. Although these effects nicely match those induced by Rhine water, it should be realized that they are rather non-specific. Finally, the laboratory studies do not support the hypothesis that the (potential) carcinogens in Rhine water (or concentrates thereof) may lead to detectable carcinogenic effects, even when tested at toxic water-sample doses. In agreement with this, field studies on native bream (*A. brama*) revealed very low tumour frequencies, and differences among these were not statistically significant. Nevertheless, the highest number of tumours was observed in Rhine fish (Table 6.5), and these are all liver tumours, as is usually observed in fish experiments with chemical carcinogens [39]. It is also interesting to note that a number of Rhine fish showed non-neoplastic hepatic lesions such as cholangiofibrosis (massive bile-duct proliferation) which may ultimately develop into neoplasms [35]. To summarize, at present no conclusive evidence is available that clearly relates the organic mutagens and carcinogens in river Rhine water to the biological effects observed in fish. Although it is likely that chemical carcinogens are responsible for the induction of at least the liver tumours, the scarcity of positive findings indicates that this does not comprise a greater threat to aquatic life than the overall burden of other toxic constituents present in the water.

Acknowledgements – The authors are indebted to Mrs. M. Verlaan-de Vries for expert technical assistance, to Dr. K. de Raat for valuable discussions and supply of the nitroductase-deficient Salmonella strains, and to Miss M. Kerkhoven for help in preparing the manuscript.

REFERENCES

[1] B. C. J. Zoeteman, *Ph.D. Thesis,* State University Utrecht, 's-Gravenhage, The Netherlands, 1978.
[2] C. F. van Kreijl and W. Slooff, *Toxic and genotoxic properties of organic compounds identified in river Rhine water,* RIVM-Report, National Institute for Public Health and Environmental Hygiene, P.O. Box 150, Leidschendam, The Netherlands, 1984, in preparation.
[3] Gezondheidsraad, *Advies inzake mutageniteit van chemische stoffen,* Directorate General of Public Health, Directorate General of Environmental Hygiene, Leidschendam, The Netherlands, 1981.
[4] B. N. Ames, J. McCann and E. Yamasaki, *Mutation Res.,* 1975, **31,** 347.
[5] A. E. Prein, G. M. Thie, G. M. Alink, C. L. M. Poels and J. H. Koeman, *Sci. Tot. Environ.,* 1978, **9,** 287.
[6] G. M. Alink, E. M. H. Frederix-Wolters, M. A. van der Gaag, J. F. J. van de Kerkhoff and C. L. M. Poels, *Mutation Res.,* 1980, **78,** 369.
[7] R. N. Hooftman and G. Vink, *Ecotoxicol. Environ. Safety,* 1981, **5,** 261.
[8] M. A. van der Gaag, G. M. Alink, P. Hack, J. C. M. van der Hoeven and J. F. J. van de Kerkhoff, *Mutation Res.,* 1983, **113,** 311.
[9] E. Yamasaki and B. N. Ames, *Proc. Natl. Acad. Sci. U.S.A.,* 1977, **74,** 3555.
[10] C. F. van Kreijl, H. J. Kool, M. de Vries, H. J. van Kranen and E. de Greef, *Sci. Total. Environ.,* 1980, **15,** 137.
[11] H. J. Kool, C. F. van Kreijl, H. J. van Kranen and E. de Greef, *Chemosphere,* 1981, **10,** 85.
[12] W. Slooff and C. F. van Kreijl, *Aquat. Toxicol.,* 1982, **2,** 89.
[13] C. L. M. Poels, *Untersuchung über die mutagenen Eigeschaften des Rheinwassers,* Paper presented at the 8th IAWR, 18–20 May 1981, Amsterdam.
[14] M. A. van der Gaag, A. Noordsij and J. P. Oranje, in *Mutagens in our Environment,* M. Sorsa and H. Vaino (eds.), Liss, New York, 1982, p. 277.
[15] F. van Hoof and J. Verheijden, *Sci. Total. Environ.,* 1981, **20,** 15.
[16] RIWA, *The quality profile of the river Meuse* (in Dutch), Rijncommissie Waterleidingbedrijven, Amsterdam, 1983.
[17] D. M. Maron and B. N. Ames, *Mutation Res.,* 1983, **113,** 173.
[18] C. F. van Kreijl, H. J. Kool, H. J. van Krane, W. Slooff and E. de Greef, *3rd Intern. Conf. Environmental Mutagens,* Tokyo, Japan, 1981, Abstr. 3P13.
[19] C. F. van Kreijl, M. Verlaan-de Bries, H. J. van Krannen, H. J. Kool and E. de Greef, *Mutation Res.,* 1983, **113,** 313.
[20] H. S. Rosenkranz and R. Mermelstein, *Mutation Res.,* 1983, **114,** 217.
[21] R. J. Spanggord, K. E. Mortelmans, A. F. Griffin and V. F. Simmon, *Environ. Mutagenesis,* 1982, **4,** 163.

[22] D. F. S. Natusch and B. A. Tomkins, *Anal. Chem.*, 1978, **50,** 1429.
[23] R. Concin, E. Burtscher and O. Bobleter, *J. Chromatog.*, 1980, **198,** 131.
[24] J. R. Bend and M. O. James, in *Biochemical and Biophysical Perspectives in Marine Biology,* D. C. Malins and J. R. Sargent (eds.), Academic Press, London, 1978, Vol. 4, p. 125.
[25] R. M. Love, *The Chemical Biology of Fishes,* Academic Press, London, 1980.
[26] C. N. Statham, M. J. Melancon and J. J. Lech, *Science,* 1976, **193,** 680.
[27] C. F. van Kreijl, A. C. van den Burg and W. Slooff, in *Mutagens in our Environment,* M. Sorsa and H. Vaino (eds.), Liss, New York, 1982, p. 287.
[28] M. A. van der Gaag, J. F. J. van de Kerkhoff, H. W. van der Klift and C. L. M. Poels, *Environ. Monit. Assess.*, 1983, **3,** 247.
[29] C. L. M. Poels and J. J. T. W. A. Strik, in *Sublethal Effects of Toxic Chemicals on Aquatic Animals,* J. H. Koeman and J. J. T. W. A. Strik (eds.), Elsevier, Amsterdam, 1975, p. 86.
[30] C. L. M. Poels, M. A. van der Gaag and J. F. J. van der Kerkhoff, *Water-Res.*, 1980, **14,** 1029.
[31] W. Slooff, D. de Zwart and J. F. J. van de Kerkhoff, *Aquat. Toxicol.,* 1983, **4,** 189.
[32] W. Slooff, C. F. van Kreijl and A. J. Baars, *Aquat. Toxicol.,* 1983, **4,** 1.
[33] W. Slooff and D. de Zwart, *Sci. Total. Environ.,* 1983, **27,** 149.
[34] W. Slooff, *Aquat. Toxicol.,* 1982, **2,** 157.
[35] W. Slooff, *Aquat. Toxicol.,* 1983, **3,** 127.
[36] H. J. Kool, C. F. van Kreijl, E. de Greef and H. J. van Kranen, *Environ. Health Perspect.,* 1982, **46,** 207.
[37] A. Crosby Longwell and J. B. Hughes, *Rapp. P.v. Rèun. Cons. Int. Explor. Mer.,* 1980, **179,** 275.
[38] J. A. Beardmore, C. J. Barker, B. Battaglia, R. J. Berry, A. Crosby Longwell, J. F. Payne and A. Rosenfield, *Rapp. P.v. Rèun. Cons. Int. Explor. Mer.,* 1980, **179,** 299.
[39] T. Matsushima and T. Sugimura, *Prog. Exp. Tumor Res.,* 1976, **20,** 367.

7

The detection of mutagens in the tissue of marine organisms exposed to environmental pollutants

J. M. PARRY, Department of Genetics, University College of Swansea, Singleton Park, Swansea, Wales

7.1 INTRODUCTION

A variety of screening systems have been developed for use with biological indicator organisms in the detection of the possible mutagenic activity of environmental chemicals. Such systems involve the use of organisms ranging from relatively simple bacteria, through mammalian cells in tissue culture, to whole animals. Most of the *in vitro* systems utilize the procedure of exposing cells of known genotype to test agents, and the plating of treated cells on a selective medium to detect those cells which have been mutated by the test agent. In a large number of cases, mutagenic activity in the screening systems has been shown to be correlated with the ability of a chemical to produce tumours either in experimental animals or humans [1].

The marine environment and inland waters can be assayed for the presence of mutagenic chemicals in two fundamentally different ways. Chemical analysis can be performed with water samples to identify constituent chemicals, which can then be individually tested for possible mutagenic activity. This often represents a formidable task because of the very large number of chemicals, both natural and man-made, that are detectable and must therefore be screened individually for mutagenic activity. However, such studies have been performed on the constituents of some samples of drinking water in the U.S.A. and have revealed the presence of mutagenic chemicals [2].

The second main approach involves the collection of water samples, concentration of the constituents, and exposure of the concentrate to a variety of mutagenic screening systems. However, this procedure has at least three fundamental limitations: the problems of developing suitable concentration procedures, the production of a test sample that reflects the chemical proportions in the original sample and, perhaps most importantly, the difficulty of detecting mutagens in the presence of toxic chemicals. For example, a mutagenic chemical might be undetectable if the test assay is performed in the presence of a chemical which is highly toxic to the test organism. Under such conditions the toxicity of one constituent may mask the activity of others. Suitable techniques for concentration and testing of chemicals are described elsewhere in this volume.

We have attempted to overcome some of the problems of the second approach by using aquatic organisms as concentrators of potentially mutagenic chemicals. By the use of such living organisms, we have also provided selectivity against the possible masking effects of toxic chemicals, on the assumption that such toxicity would lead to death of the organisms used. Extracts of the tissues of such organisms are presumed to represent a sample of potentially mutagenic chemicals in the tissues exposed. Most of our work has involved the development of an assay system based on the extraction of tissues derived from the edible marine mussel, *Mytilus edulis,* and the screening of these extracts for mutagenic activity, with a number of indicator species capable of detecting the induction of genetic damage. By using such techniques, we have been able to identify the presence of chemicals capable of inducing mutation in a number of test systems.

The techniques described here have been used by us [3] to demonstrate the presence of mutagens in the body tissues of marine species. However, there is no theoretical reason why they should not be applied to terrestrial species.

7.2 PREPARATION OF TISSUE SAMPLES

Because of the variety of chemicals encountered in the aquatic environment which may accumulate in body tissues of aquatic organisms, it is unlikely that a single solvent procedure would result in the efficient extraction of all of them. Ideally some consideration should be made of the possible chemical nature of the agents present in a specific environment and the extraction procedure modified appropriately. When such information is unavailable then the investigator is faced with the need to compromise, and a range of solvent procedures should be used. In this chapter, three procedures are described, which use different solvents that have been shown to be of considerable practical value in the extraction of potentially genotoxic chemicals in the tissues of marine species.

The techniques have been used with the tissues of the edible mussel, *Mytilus edulis,* but may be applied to other species.

Ethanol extraction

1. Collect samples and transfer to the laboratory in an ice-box within 3 hr.
2. Open shells, drain shell liquid, and rinse briefly in distilled water to remove materials such as sand.
3. Remove and chop up tissues, add 50 g of tissue to 100 ml of absolute ethanol, and homogenize in a Waring Blender until the tissue is dispersed.
4. Transfer to a glass container and leave overnight at 4°C.

5. Centrifuge and collect the supernatant liquid.
6. Evaporate to dryness in a rotary evaporator.
7. Re-suspend in 10 ml of a solvent appropriate for use in the mutagenicity test (dimethylsulphoxide is a most convenient solvent).
8. Filter and sterilize the extract, and store it at –20°C until required.

Chloroform/methanol extraction

1. Remove and chop 50 g of tissue, add to 1 litre of chloroform/methanol mixture (2:1 v/v), and homogenize in a Waring Blender until the tissue is dispersed.
2. Transfer to a glass container and leave overnight at 4°C.
3. Partition solvent mixture with 0.05*M* potassium chloride.
4. Collect the organic layer and add anhydrous sodium sulphate to it.
5. Filter.
6. Evaporate the filtrate to dryness in a rotary evaporator.
7. Re-suspend in 10 ml of an appropriate solvent.
8. Filter and sterilize the extract and store it at –20°C until required.

Acid digestion

1. Remove and chop 15 g of tissue and place in a round-bottomed flask.
2. Add 20 ml of a 5:1 v/v mixture of nitric acid and 60% perchloric acid.
3. Boil the mixture under reflux until all tissues are seen to have dissolved (generally about 15 min).
4. Allow mixture to cool, neutralize to pH 7.0 with sodium hydroxide and then make up to 50 ml with distilled water.
5. Sterilize extract by passing it through a 0.45-μm membrane filter.

Control preparations may be prepared by performing the extractions with organisms collected from previously characterized ‘clean’ sites or by using organisms ‘conditioned’ in the laboratory. In our experience organisms such as shellfish rapidly excrete their body burden of mutagenically active pollutants if they are placed in fresh changes of sea-water over a period of 1–3 weeks.

7.3 DETECTION OF NUTRILITES

Most microbial mutagenicity assays involve the measurement of the induction of prototrophic colonies in an auxotrophic strain. For example *Escherichia coli* WP_2 assays involve the production of tryptophan-independent colonies whereas *Salmonella typhimurium TA* assays involve the production of histidine-independent colonies. In all such assays, cells are grown in or on a selective medium containing none or only a limited amount of the specific nutrient (nutrilite) required by the strain under test. In these experiments the presence of nutrilites such as tryptophan and histidine in the test extract may lead to additional periods of cell growth during which spontaneous mutants may arise, producing artificially ‘positive’ results.

To avoid the production of artefacts in the assays, because of the presence of nutrilites, it is therefore an absolute necessity that tissue extracts are tested for the presence of factors capable of supplementing growth. If such factors are present then mutagenicity test procedures must be used which are not influenced by their presence, such as the assay of drug-resistant colonies rather than of prototrophs.

A number of techniques have been developed for the detection of nutrilites and examples are given below.

(A) If extracts have been tested for genetic activity by plate assays as described in Section 7.4 (B), below, any increased cell growth due to the presence of nutrilites may be assayed directly by sampling representative regions of selected plates.

1. Using a sterile glass tube (the end of a Pasteur pipette is suitable), remove 10 cores of agar, 9 mm in diameter, from areas of the plate showing background growth but no large prototrophic colonies.
2. Add the discs to 20 ml of sterile saline and suspend the cells (from control plates and treated plates) by vigorous or ultrasonic agitation.
3. Dilute the suspensions appropriately and plate them on suitable nutrient media.
4. Incubate the plates overnight at 37°C for bacteria or for 3 days at 28°C for yeast.
5. Compare the cell titres of the control and treated plates and determine whether any growth supplementation has taken place.

(B) If extracts are to be used in a fluctuation test [Section 7.4 (A)] then there are a number of techniques which can successfully be used to assay for the presence of nutrilities.

(B1) – Macroscale fluctuation test

1. For each extract to be tested, prepare three sterile 25-ml 'Universal' bottles containing 2-ml aliquots of cell culture, medium and tissue extract, in the same proportion as used in the bacterial fluctuation test described in Section 7.4 (A).
2. Incubate the cultures for 18 hr at 37°C (for bacteria) or 28°C (for yeast).
3. Dilute the cultures as appropriate and plate on three replicate nutrient plates and incubate overnight (for bacteria) or for 3 days (for yeast).
4. Compare the colony counts of cultures containing extracts with those of control cultures to determine whether increased cell growth has occurred, indicating the presence of nutrilites.

(B2) – Microscale fluctuation test

1. Prepare a mixture of cell suspension, Bromocresol Purple indicator, and medium in the same proportions as used in the fluctuation test described in Section 7.4 (A), but without growth supplement.
2. Dispense 200 μl of cells and medium into the 96 wells of a microtitre plate.

3. Divide the wells into rows running from left to right, and columns running from top to bottom.
4. To the wells in a given row add a fixed concentration of supplement, increasing the concentrations (e.g. 0.2, 0.4, 0.6, 0.8, 1.0 μg/ml of histidine per well) from top to bottom of the plate.
5. To the columns add 5 μl of extract per well, using one column of wells for each extract to be tested and one for a control series.
6. Incubate the plates for 3 days at 37°C in a humidified incubator.
7. During the incubation period the indicator will change from purple to yellow as the medium becomes acidified by the increased cell growth.
8. Comparison of the individual columns will identify any extract in which growth has occurred at lower levels of supplementation.
9. Extracts producing such colour changes at low concentrations of supplement can be considered to contain nutrilites.

(B3) – Post-treatment assay for nutrilites

In a fluctuation test for mutagenicity, comparisons are made between 'positive' wells or tubes in which the growth of prototrophs has occurred, with 'negative' wells or tubes in which growth of auxotroph has occurred only to an extent appropriate to the level of supplement supplied. The level of growth occurring in these negative wells can be used as a measure of any growth supplementation produced by test extracts.

1. After the completion of a fluctuation test identify samples of 10 'negative' wells in both the control and the treated series.
2. Using a sterile Pasteur pipette gently agitate the contents of a well to suspend all the cells, and transfer the contents to a sterile 25-ml 'Universal' container.
3. Dilute appropriately with sterile saline and plate three replicates per sample.
4. Incubate overnight at 37°C (for bacteria) and for 3 days at 28°C (for yeast).
5. Compare the colony numbers obtained from the control and treated samples to identify any 'negative' wells in which additional growth has taken place. The median number of cells per well, calculated from the colony counts, is an appropriate statistic with which to compare the control and treated series and to identify the presence of nutrilites in tissue extracts.

7.4 DETECTION OF MUTAGENIC ACTIVITY

In the preparation of suitable tissue extracts the experimenter is faced with selection of the most appropriate mutagenicity tests to utilize. This decision is essentially dependent on the expertise and facilities available in the laboratory concerned. Some of the techniques that have been successfully employed in our laboratory, together with some examples of the results obtained, are given below.

(A) Bacterial Mutagenicity Tests

The most extensively validated bacterial mutagenicity test is the agar-overlay technique developed by Ames *et al.* [4]. This technique has been extensively described in the literature and readers are referred to a variety of comprehensive publications describing its use (for example Maron and Ames [5]). The assay involves the combination, on an agar plate, of bacterial cultures auxotrophic for histidine with a test material in the presence of a supplementary activation system derived from rat liver ('S9 mix'). One of the primary advantages of the assay is that the S9 mix has an active lifetime of many hours when suspended in agar. The agar-overlay technique can be used with bacteria to assay tissue extracts. However, in our experience it lacks the sensitivity of the bacterial fluctuation test in detecting the relatively low levels of mutagenic chemicals found in tissue extracts.

The bacterial fluctuation test, modified for mutagenicity assays by Green *et al.* [6], is a liquid assay of high sensitivity, particularly when used as the microtitre-plate method developed by Gatehouse [7]. A suitable modification of this technique is as follows.

1. Prepare 20 ml of Davis-Mingioli salts solution, containing 4 μg of biotin and 0.16 g of glucose, and inoculate with 5×10^5 bacterial cells/ml.
2. Add the appropriate amount of extract to be tested, as determined by a pretest toxicity assay.
3. Dispense this solution into 96-well microtitre plates at 200 μl/well (1 plate per extract concentration).
4. Incubate the plates at 37°C and score on the third and fourth days. Wells in which growth has occurred contained prototrophic revertants. The ease of reading the microtitre plates is improved if the indicator Bromocresol Purple is added to the medium. With this indicator, wells in which growth has occurred undergo a colour change from purple to yellow because of the acidification caused by the presence of prototrophs.

The appropriate statistic for analysis of the data is:

$$\chi^2 = \frac{2n(t - c - \frac{1}{2})^2}{(t + c)(2n - t - c)} \quad ; \text{ 1 degree of freedom}$$

where n = number of wells in a microtitre plate; t = number of positive wells in a treated plate; c = number of positive wells in a control plate.

To perform the fluctuation test in the presence of an exogenous activation mix, the modification described by Gatehouse and Delow [8] may be used.

1. Prepare 4 ml of Davis-Mingioli salt solution containing 0.4% glucose, 10 μg/ml biotin, 1.5 μg/ml histidine and 30 μl of 3-hr bacterial culture (giving approximately 4×10^6 bacteria/ml).

2. Add 1 ml of 'S9 mix' consisting of 0.1 ml of S9 fraction (30 mg/ml), 8 μmoles of $MgCl_2$, 33 μmoles of KCl, 5 μmoles of glucose-6-phosphate, 4 μmoles of NADP and 100 μmoles of sodium phosphate, adjusted to pH 7.4.
3. Add positive control mutagen or tissue extract in 0.5 ml of Davis-Mingioli salt solution to give a final volume of 5.5 ml.

Dispense the mixture into a 96-well microtitre plate (50 μl per well). Fit lids on the trays and incubate overnight at 37°C in a sealed container providing a humid atmosphere.

On the following day top up each well in the tray with 150 μl of Davis-Mingioli salt solution (single strength) containing 0.8% glucose and 5 μg/ml Bromocresol Purple. Incubate the plates for a further 3 days at 37°C before scoring.

(B) Yeast Agar-overlay Assay

1. Suspend stationary-phase cells (of an appropriate tester strain) in saline, 10^7 cells/ml.
2. Mix together test extract (up to 4%) with yeast minimal medium kept molten at 45°C, together with an appropriate number of yeast cells. For yeast strain *JD1* we have found that 35×10^3 cells/ml and 35×10^2 cells/ml are appropriate for the detection of histidine-independent and tryptophan-independent prototrophs respectively. If required, an exogenous S9 mix may also be added.
3. The minimal medium used is supplemented with a sufficient concentration of the specific auxotrophic requirement of the strain used, to allow three cell-divisions to occur during incubation in the presence of the tissue extract. In the case of *JD1* this is 20 μg/ml tryptophan and 0.1 μg/ml histidine for the detection of histidine-independent prototrophs, and 20 μg/ml histidine and 0.1 μg/ml tryptophan for the detection of tryptophan-independent prototrophs.
4. Incubate the plates in the dark at 28°C for 9 days before scoring the frequencies of prototrophic colonies.

(C) In vitro *Cytogenetic Assay*

Tissue extracts may be assayed for their ability to induce chromosome aberrations in various cultured cell lines. We have found that the rat liver line RL_4 developed by Dean and Hodson-Walker [9] is particularly convenient for this purpose. For such studies particular consideration must be given to the possible toxicity of solvents used in the extraction procedure. Dimethylsulphoxide has proved suitable in our laboratory, but alternatives should be considered. We recommend that samples should be tested up to toxicity levels of 50% growth-inhibition in at least two separate experiments.

7.5 EXAMPLES OF THE USE OF THE TISSUE BIOASSAY

The mutagenicity bioassay of the tissues of aquatic organisms has the potential to provide information on the activity and distribution of genotoxic chemicals present in any aquatic environment.

The first question that might be posed by the genetic toxicologist dealing with an aquatic environment is whether genetically active chemicals are present at detectable levels. An answer can readily be provided by the use of microbial indicator species and tissue extracts from a suitable aquatic species present at a variety of sites. Table 7.1 illustrates some of the data obtained [10] by testing (in yeast) whole tissue extracts of the edible mussel *Mytilus edulis* collected from a variety of sites. The results indicate the presence of mutagenic chemicals in the tissue of mussels collected from only some of the sites.

The species selected for screening purposes will depend on a variety of factors, such as population size and ease of collection and handling. An example of the assay of a variety of shellfish at a single site [10] is provided in Table 7.2. The results show that at a site characterized as containing mutagens, various species are suitable biological monitors and can be used in the assay.

For any genotoxic agent a primary consideration is ability to reach target organs such as the sites of primary tumour formation. An example [11] of a study of tissue distribution of mutagenic activity in the edible mussel is shown in Table 7.3. It shows that maximum levels of activity are to be found in the mantle tissue of mussels collected from a particular sampling site.

In any monitoring study of the presence of mutagens in aquatic organisms consideration must be given to factors such as (a) season of collection, as at some sites maximum mutagenic activity may be detectable only at times of high metabolic activity in the species sampled [10] and (b) local pollution incidents, depending on the nature of the pollutant involved. Particular sites may show both increases and decreases in the activity of mutagens in tissue extracts.

Once mutagenic activity has been demonstrated in tissue bioassays by using microbial species, an important question is then whether the microbial response can be reproduced in higher organisms. The type of 'verification' assay to be performed will depend on available expertise and possible legislative requirements. In our experience an *in vitro* cytogenetic assay for the induction of chromosomal aberrations in cultured mammalian cells is suitable for such verification purposes. Table 7.4 illustrates the results of the assay of a chloroform/methanol extract of mussel tissue, with cultured rat liver RL_4 cells [10]. The data presented demonstrate the induction of aberrations in this cell line by chemicals extracted from mussel tissue. Further verification of activity can be obtained by using the extensive range of mutagenicity test assays available, which will provide data on activity both *in vivo* and *in vitro* [12].

The specific chemicals producing the mutagenic response observed in tissue bioassays can be identified by chemical analysis combined with further mutagenicity studies. The microbial assay systems provide convenient techniques for the verification of activity in specific chemical fractions and chromatographic bands.

Table 7.1 – The effects of alcohol extracts of the edible mussel *Mytilus edulis* collected from a variety of marine sites, on the induction of mitotic gene conversion in the yeast *Saccharomyces cerevisiae* [10]

Concentration of extract/plate	% Viability of treated yeast culture	His^+ prototrophs per 10^6 survivors ± S.D.*	Trp^+ prototrophs per 10^6 survivors ± S.D.*
Control 1	91.4	26.8 ± 4.6	23.1 ± 4.3
4% alcohol control 1	73.1	25.6 ± 5.0	37.0 ± 6.0
4% alcohol control 2	101.4	11.8 ± 3.0	21.4 ± 3.9
Plymouth extract 4%	95.4	493.6 ± 21.3	463.5 ± 18.6
Mumbles extract 4%	62.9	323.6 ± 19.2	387.7 ± 20.6
Caswell Bay 4% extract	71.4	276.0 ± 16.6	276.8 ± 16.6
Anglesey 4% extract	73.4	18.3 ± 4.2	29.6 ± 5.4
EMS (1 μg/ml)	90.3	900.9 ± 26.7	1095.0 ± 29.0
Control 1	98.3	62.2 ± 8.4	33.1 ± 33.1
Swansea (St. Helens) 4%	101.3	256.7 ± 12.3	192.6 ± 11.6
Solva (S. Wales) 4%	104.1	43.2 ± 9.1	23.9 ± 8.4
Milford Haven (S. Wales) 4%	87.3	175.4 ± 12.1	200.4 ± 15.6
Llangranog (S. Wales) 4%	83.4	284.7 ± 11.9	390.9 ± 19.7
4% alcohol control 1	75.6	65.4 ± 8.4	22.3 ± 4.3
4% alcohol control 2	97.6	47.3 ± 7.1	21.1 ± 3.1
EMS (1 μg/ml)	86.2	1050.7 ± 29.6	712.3 ± 27.0

*S.D. = standard deviation.

Table 7.2 – The induction of mitotic gene conversion in yeast strain *JD1*, produced by alcohol extracts of the tissue of 3 species of shell fish collected from Mumbles, W. Glamorgan, Wales [10]

Concentration and extract*	% Viability of treated yeast culture	His^+ prototrophs/10^6 survivors ± S.D.†	Trp^+ prototrophs/10^6 survivors ± S.D.†
Control	101.2	34.2 ± 5.6	11.4 ± 2.7
Control 2	97.6	29.1 ± 7.1	13.6 ± 3.1
1% Ethanol control	104.9	31.0 ± 4.8	10.1 ± 2.5
1% Ethanol control	103.4	28.7 ± 3.7	9.4 ± 2.3
1% Extract of *P. vulgata* 1	81.3	51.2 ± 7.2	41.3 ± 5.9
1% Extract of *P. vulgata* 2	92.8	74.1 ± 6.3	50.7 ± 4.8
1% Extract of *M. edulis* 1	78.6	219.1 ± 11.2	113.3 ± 15.2
1% Extract of *M. edulis* 2	85.2	176.4 ± 19.1	146.7 ± 13.7
1% Extract of *N. lapillus* 1	90.1	287.6 ± 21.3	189.9 ± 28.4
1% Extract of *N. lapillus* 2	72.7	315.2 ± 24.9	172.1 ± 19.7
1 μg/ml EMS	67.4	513.1 ± 47.2	316.7 ± 21.4

*Extract prepared from 50 g of whole tissue in 100 ml of ethanol.

†S.D. = standard deviation.

Table 7.3 – The effects of alcohol extracts of a variety of tissues of the edible mussel *Mytilus edulis* on the induction of mitotic gene conversion in the yeast strain *JD1* [11]

Treatment	Viability	Trp^+ prototrophs per 10^5 survivors ± S.D.*	His^+ prototrophs per 10^6 survivors ± S.D.*
Control	100.0	19.1 ± 1.8	36.7 ± 2.3
Negative control 2% ethanol	101.3	22.4 ± 1.7	32.4 ± 1.9
Muscle 1%	97.3	16.6 ± 1.4	37.5 ± 2.3
2%	94.4	29.3 ± 1.8	41.0 ± 2.9
Mantle 1%	101.9	94.5 ± 3.9	170.3 ± 11.6
2%	93.2	135.7 ± 8.2	589.6 ± 23.4
Hepatopancreas 2%	85.3	27.9 ± 1.5	41.6 ± 2.7
Foot 2%	92.4	17.6 ± 0.9	39.4 ± 2.1
Reproductive system 2%	76.8	14.3 ± 2.3	20.7 ± 1.9
2 μg/ml ethyl methanesulphate	62.4	216.7 ± 19.8	296.3 ± 28.4

*S.D. = standard deviation.

Table 7.4 – The induction of chromosome aberrations in rat cell culture line RL_4 produced by exposure to a chloroform/methanol extract of whole tissue of the edible mussel *Mytilus edulis* [10]

Exp.	Concentration of mussel extract (g/ml)	% Cells with aberrations	Chromatid			Chromosomal	
			Gaps	Breaks	Interchanges	Gaps	Others
	0	3	2	1	–	–	–
2	0.25	7	5	2	–	–	–
	0.5	8	5	4	–	–	–
	1.0	12	6	6	–	2	1
	1.5	16	6	7	–	3	2
	2.0	16	8	3	2	1	3
	5.0	24	12	7	1	4	1
	10.0	18	8	8	–	2	1
1	20.0	20	10	4	–	3	5
	40.0	17	10	3	–	1	3
	80.0*	15	6	6	–	1	2
	Positive control†	66	60	32	10	–	–

*100 metaphases scored at each concentration.

†5 μg/ml 7,12-dimethylbenz(*a*)anthracene (50 metaphases).

REFERENCES

[1] J. McCann, E. Choi, E. Yamasaki and B. N. Ames, *Proc. Natl. Acad. Sci. U.S.A.*, 1975, **72**, 5135.
[2] V. F. Simmons, K. Kaukanen and R. G. Tarditt, in *Progress in Genetic Toxicology*, D. Scott, B. A. Bridges and F. H. Sobels (eds.), Elsevier/North Holland, Amsterdam, 1977, pp. 249–258.
[3] J. M. Parry, D. Tweats and M. A. J. Al-Mossawi, *Nature*, 1976, **264**, 538.
[4] B. N. Ames, W. E. Durston, E. Yamasaki and F. D. Lee, *Proc. Natl. Acad. Sci. U.S.A.*, 1973, **70**, 2281.
[5] D. M. Maron and B. N. Ames, in *Handbook of Mutagenicity Test Procedures*, 2nd Ed., B. J. Kilney, M. Legater, W. Nichols and C. Ramel (eds.), Elsevier/North Holland, Amsterdam, 1984, pp. 93–140.
[6] M. H. L. Green, W. J. Muriel and B. A. Bridges, *Mutation Res.*, 1976, **38**, 33.
[7] D. Gatehouse, *Mutation Res.*, 1978, **53**, 289.
[8] D. Gatehouse and G. F. Delow, *Mutation Res.*, 1979, **60**, 239.
[9] B. J. Dean and G. Hodson-Walker, *Mutation Res.*, 1979, **64**, 329.
[10] J. M. Parry, M. Kadhim, W. Barnes and N. Danford, in *Phyletic Approaches to Cancer*, C. J. Dave (ed.), Japan Sci. Pres, Tokyo, 1981, pp. 141–146.
[11] J. M. Parry, W. Barnes and M. Kadhim, in *Progress in Environmental Mutagenesis*, Elsevier/North Holland, Amsterdam, 1980, pp. 227–248.
[12] S. Venitt and J. M. Parry, *Mutagenicity Testing, A Practical Approach*, IRL Press, Oxford, 1985.

8

Detection of genetic activity in effluents from pulp and paper mills: mutagenicity in *Saccharomyces cerevisiae*

E. R. NESTMANN, Mutagenesis Section, Environmental Health Directorate, Department of National Health and Welfare, Ottawa, Ontario, Canada

8.1 DETECTION OF GENETIC ACTIVITY

The Salmonella reverse mutation assay [1] is the most commonly used test in the field of genetic toxicology. Additional bacterial assays have been developed which use other species (e.g. *Escherichia coli*), other methods (e.g. the fluctuation test), and other end points (e.g. forward mutation, recombination, DNA damage gene duplication, or lysogenic or SOS induction). Many of these techniques have arisen as research curiosities not to be seriously considered for widespread use as screening tests. Most investigators feel that only one bacterial assay should be included in a test battery, and that the Salmonella assay is the method of choice owing to its common usage and continual development and validation.

Realization that many carcinogens may require chemical changes by mammalian metabolic enzymes before they are active [2] has led to routine *in vitro* testing procedures with a mammalian activating system [3]. Usually a homogenate of liver (S9) or liver microsomes is used as a source of these enzymes, with rats as the most common, but not exclusive, source of tissue. The incorporation of such material in mutagenicity tests is believed to mimic a degree of mammalian metabolism *in vitro.* In no way can it be claimed that the *in vivo* condition is duplicated, because many additional factors, such as uptake, transport, deactivation, tissue specificity, are introduced. For these reasons, genetic activity may be detected in many mutagenicity tests, but mutagenic potential in man must be evaluated at present by *in vivo* mammalian experiments that will allow quantitative estimates or risk.

Even with the metabolic activation capacity of S9, no single *in vitro* mutagenicity test is capable of detecting all mutagens. One possible and practical strategy used is to combine two or more assays in what is referred to as a battery [4]. Test batteries typically consist of a microbial screening assay, such as the Salmonella/mammalian-microsome test, an *in vitro* mammalian cell culture study, and a mammalian *in vivo* assay. Different tests can be arranged in many combinations, depending, for example, on the material to be tested, the expertise available, and the purpose for which the results are to be used. For those without access to mammals or tissue culture, a battery of microbial tests has been proposed [5], which combines the sensitivity and convenience of bacteria and yeast as test organisms.

The choice of tests depends on several factors, including availability of appropriate tests and relative cost. Chemical class also helps to determine which tests are suitable, since it is known that certain biological or biochemical parameters render certain systems insensitive to particular chemical classes. For example, owing to the extensive bleaching of wood products during the pulping process, pulp and paper-mill effluents contain many chlorinated compounds that are thought to be responsible for much of the genetic activity shown by these complex environmental mixtures. It is well known that the Salmonella assay is not sensitive in the detection of the mutagenicity of certain chlorinated hydrocarbons that show mutagenic and carcinogenic activity in other test systems [6]. Therefore, yeast assays may be more suitable tests for the detection of mutagenic activity in pulp and paper-mill effluents.

Thanks to the efforts of Zimmerman *et al.* [7], strains of *Saccharomyces cerevisiae* have been specially bred to allow the detection of a variety of genetic end points, namely, reverse mutation, gene conversion, aneuploidy, mitotic recombination, and other non-specific types of chromosome damage. In addition, Callen and co-workers, [8, 9] have shown that growing cells of *S. cerevisiae* produce the enzymes necessary to activate many promutagens that would not be mutagenic in *in vitro* assays without S9. This property of yeast has been exploited further in the laboratories of Parry and von Borstel. In addition, a differential survival assay using *S. cerevisiae* and *Schizosaccharomyces pombe* as an indicator test for mutagenicity has been devised to take advantage of these yeasts' ability to activate promutagens [10].

The foregoing is not meant to suggest that testing with yeasts can replace a battery of microbial and mammalian assays. However, yeasts are easily manipulated experimentally, with a minimum of equipment. In addition, the expense associated with the preparation, storage, and use of S9 can be eliminated without sacrificing the benefit of a degree of metabolic-activation capacity. These features make yeast very appropriate for work in any testing laboratory, but perhaps even more so, for example, in developing countries, owing to cost and availability of equipment.

The utility of inclusion of a yeast assay in a battery of microbial tests has been demonstrated repeatedly with various agents. For example, studies of 62 chemical constituents of pulp and paper-mill effluent [11, 12] compared two yeast tests which detected either gene conversion [7] or reverse mutation [13].

Seventeen compounds were mutagenic in yeast, whereas only twelve had been positive in *Salmonella*, providing additional support for the value of yeast as an organism for mutagenicity testing. This chapter describes and discusses the studies performed to date on the mutagenicity of pulp and paper-mill effluent in yeast.

8.2 PULP AND PAPER-MILL EFFLUENT – A COMPLEX MIXTURE

Effluents from pulp and paper mills consist of a complex mixture of dissolved lignin and cellulose degradation products, other wood extracts, and chlorinated organic compounds from pulping processes. Not only are these effluents toxic to fish in the receiving waters, but many of the constituent compounds are not readily degraded by natural means. These persistent compounds are of concern for the environment, owing to their deposition and bioaccumulation. Since the receiving waters may be used as sources of drinking water, the impact of constituents of these effluents on water supplies is also a potential problem for human health.

Approaches to the evaluation of mixtures (such as pulp-mill effluents) for mutagenic potential fall into two general categories: the study of the mixture itself or the study of the mixture's individual components. At the onset, it seems more practical and relevant to direct experimental efforts to study of the mixtures themselves, since they are the materials released into the environment and those to which organisms would seem to be exposed. This approach involves collection of effluents, receiving waters, or samples downstream, and testing for mutagenic potential. Depending on the levels of contaminants in the test materials, the wastewater may be concentrated by extraction or by separation procedures such as column chromatography or reverse osmosis.

8.3 YEAST TESTING OF AN EXTRACT OF EFFLUENT

An initial study using yeast tests for detecting potential genetic activity of samples of pulp and paper-mill effluent involved the study of the first chlorination-stage effluent from a bleached kraft pulp mill. Such samples had been shown previously to be mutagenic in bacteria and Chinese hamster ovary cells [14, 15]. Although the effluent itself had previously been found to be mutagenic [14, 15], the organic compounds in the sample used in the yeast study [16] were concentrated by XAD-2 affinity chromatography to reduce the volumes required for testing. Strain D7 of yeast [7] was chosen for testing the concentrate from the XAD-2 columns. This strain enables the detection of several genetic end points, including gene conversion and mitotic recombination, results for which are shown in Table 8.1. It is apparent from Table 8.1 that a dose-related increase in recombinant (twin-spot) colonies was induced by the extract of the chlorination-stage effluent. The frequency of recombinants compared to that for the solvent control was 4 times higher at the lowest dose

Table 8.1 – Induction of mitotic recombinants and gene convertants in yeast strain D7 with effluent extract

Test material	Dose (mg/ml)	Survival (%)[a]	Aberrant colonies per 10^3 survivors*	Twin-spots per 10^3 survivors*	Convertants per 10^6 survivors†
Water		100	0.14	0.05	7.34
DMSO		90.3	0.16	0.05	10.8
Extract	0.5	48.7	0.99	0.20	9.36
	0.7	37.3	1.03	0.39	35.5
	1.0	14.2	1.12	0.31	39.0
	2.0	2.0	9.52	0	–
Ethyl methanesulphonate	1.18	78.0	14.6	2.96	112

*Colonies scored on YEPD medium (0.2 ml/plate), 40 plates per dose, except for the positive control (10 plates).
†Counted on tryptophan-omission medium in quintuplicate (0.2 ml/plate).

tested, and 8-fold higher at its maximum. The frequency declined with decreasing numbers of surviving recombinant (and normal) colonies. The frequency of aberrant colonies other than twin-spots (due to non-specific genetic events, including DNA damage, aneuploidy and gene conversion) was about the same for all but the highest treatment dose, 7-fold higher than that for the control. The 60-fold increase at the highest dose may be exaggerated, in that few colonies were available for an accurate determination of survival. Table 8.1 also shows that the frequency of convertants at the lowest test dose was about the same as for the control, but it increased by 300–400% at the intermediate doses, with no surviving convertants at the highest dose. Reversion of an isoleucine auxotrophic mutation in strain D7 was not examined. The ability of this extract to induce reverse mutation in the *Salmonella* plate test [15] confirmed previous results [15]. The bacterial results and the effects of a mammalian activating preparation (S9) are provided in detail by Kamra *et al.* [16].

8.4 YEAST TESTING OF COMPONENTS OF EFFLUENTS

Yeast tests have been an integral part of a comprehensive programme [17] to identify individual mutagens that are present in pulp and paper-mill effluents. Two studies have been performed, involving 62 compounds, with diploid strain D7 to detect gene conversion and haploid XV186-14C to detect reverse mutation [11, 12]. The chemicals tested in the two yeast studies that are positive either in yeast or in bacteria are listed in Table 8.2. It is clear that a larger proportion (17/20) of these chemicals were mutagenic in yeast than in bacteria (10/18; 12/18 when a test modification was employed). The literature contains additional reports of mutagens detected in pulp and paper-mill effluents, and a summary of these positive findings is shown in Table 8.3. Many more of these chemicals have been tested in bacteria than in yeast, and the proportions of mutagens in the tests are 7/14 for bacteria (10/14 with a test modification) and 4/5 for yeast. Combining the findings of Tables 8.2 and 8.3 provides a comparison of the effectiveness of yeast and bacterial assays in detection of mutagens in pulp and paper-mill effluent. Thus 84% (21/25) of the mutagens found were mutagenic in yeast and 53% (17/32); 67% or 22/32 with a test modification) were active in bacteria, providing evidence that testing with multiple end points (reverse mutation, gene conversion and mitotic recombination) in a eukaryote is more sensitive for this group of compounds than the Salmonella reverse mutation test. Since only yeast and bacterial *mutagens* are listed in Tables 8.2 and 8.3, of course all the chemicals were found to be active in one organism or the other, showing that the tests complement one another in the detection of mutagens among the chemicals screened. These studies, therefore, provide additional support for the value of using both yeast and bacteria in a battery of microbial assays in mutagenicity testing [5, 24]. The superior detection by yeast assays of the mutagens in the studies reviewed here, compared with bacterial assays, may be due in part to the lack of sensitivity of the Salmonella tester strains to chlorinated compounds [6].

Table 8.2 – Summary of mutagens identified in pulp and paper mill effluents that have been tested in yeast [11, 12].

Chemical	Response in yeast strains* D7	XV185-14C	Response in bacteria*
acetovanillone [11]	–	+	–[18]
3-chloro-*cis*-muconic acid [11]	–	+	–[18]
4,5-dichlorocatechol [11]	–	+	–[18]
4,5-dichloroguaiacol [11]	–	+	–[18]
neoabietic acid [11]	–	+	+[19]
7-oxodehydroabietic acid [11]	–	+	–[19]
1,1,2,3-tetrachloropropene [11]	–	+	+[18]
1,1,2,3,3-pentachloropropene [11]	–	+	+[18]
1,3-dichloroacetone [12]	+	w	+[20]
1,1,1-trichloroacetone [12]	+	–	NT
1,1,3-trichloroacetone [12]	+	w	+[17]
1,1,1,3-tetrachloroacetone [12]	w	+	NT
1,1,3,3-tetrachloroacetone [12]	+	w	+[17]
pentachloroacetone [12]	+	–	+[17]
hexachloroacetone [12]	+	–	+[17]
dibromochloromethane [12]	+	–	+[21]†
bromodichloromethane [12]	w	w	+[21]†
trichloroacetaldehyde [12]	–	+	–[20]
1,1,2,2-tetrachloroethane [11]	+[9]; –[11]	–	+[22]; –[18]
trans-anethole [11]	–	–	+[22]; –[18]

*Responses are designated by the following symbols: +, positive, –, negative, w, weakly positive, NT, not tested.
†Desiccator used as an exposure chamber.

Table 8.3 – Mutagens reported to be in pulp and paper-mill effluent.

Chemical	Response in yeast*	Response in bacteria*
1,2-dichloroethane	NT	+[18]†
dichloromethane	NT	+[18, 24]†
1,1,1-trichloroethane	–[25]	+[18, 26]†
benzyl chloride	NT	+[27]
bromo-*p*-cymene	NT	+[28]
dichloro-*p*-cymene	NT	+[28]
safrole	+[29]	–[27]
chloroform	+[9]	–[18, 27]
carbon tetrachloride	+[9]	–[27]
2-chloropropenal	NT	+[20]
3-chloro-4-dichloromethyl-5-hydroxy-2-(5H)-furanone	NT	+[30]
trichloroethylene	[31]§	+[21]§
tetrachloroethylene	NT	+[32]
monochloroacetaldehyde	NT	+

*Responses designated as in Table 8.2.
†Response found when desiccators were used as exposure chambers.
§The mutagenicity of this compound is in doubt owing to the presence of stabilizers in the sources for testing.

8.5 FURTHER DEVELOPMENT OF YEAST FOR MUTAGENICITY TESTING

Despite the favourable comparison above, that shows yeast assay to be the test of choice for identifying mutagens in pulp and paper-mill effluent, and other examples in the literature, there remains some scepticism about its sensitivity and the validity of including this eukaryotic micro-organism in batteries of mutagenicity tests. Ames and co-workers have devoted a great deal of effort to developing and validating the Salmonella/mammalian-microsome assay [1]. In contrast to the efforts made with Salmonella, yeast testing has not been the subject of such a methodical, comprehensive and concerted programme. However, von Borstel, Parry, Zimmermann and their colleagues have made, and continue to make, valuable contributions to the discipline. An example of work under way to improve yeast strains for enhanced sensitivity is the recent introduction of bacterial mutator genes into yeast. Much of the success of the Salmonella assay has been attributed to the inclusion of these plasmid-bearing mutator genes (*mucA, B*) in certain tester strains. By use of a bacterial/yeast shuttle vector, these genes were shown to enhance spontaneous and ultraviolet-induced mutagenesis and survival after ultraviolet irradiation, in a yeast host [33]. These responses are the same as those found in bacterial cells, though the levels of the effects are not as pronounced in yeast. Experiments are continuing to introduce proper yeast regulatory signals to boost expression of the bacterial mutator genes in yeast cells.

Acknowledgements – I thank Dr. R. C. von Borstel and Dr. F. K. Zimmermann for encouragement with experimental work and with this contribution, respectively, Drs. A. F. Hanham and A. A. Potter for their comments and suggestions concerning the manuscript, and Cathy Gilchrist and Kathy Nesbitt for their unfailing efforts to type and edit the typescript.

REFERENCES

[1] D. M. Maron and B. N. Ames, *Mutation Res.*, 1983, **113**, 173.
[2] J. A. Miller and E. C. Miller, *Prog. Exp. Tumor Res.*, 1969, **11**, 273.
[3] H. V. Malling, *Mutation Res.*, 1971, **13**, 425.
[4] B. A. Bridges, *Nature*, 1976, **261**, 195.
[5] E. R. Nestmann, T. I. Matula, G. R. Douglas, K. C. Bora and D. J. Kowbel, *Mutation Res.*, 1979, **66**, 357.
[6] L. Haroun and B. N. Ames, in *Short-Term Tests for Chemical Carcinogens*, H. F. Stich and R. H. C. San (eds.), Springer-Verlag, New York, 1981, p. 108.
[7] F. K. Zimmermann, R. Kern and H. Rasenberger, *Mutation Res.*, 1975, **28**, 381.
[8] D. F. Callen, *Mutation Res.*, 1978, **55**, 153.
[9] D. F. Callen, C. R. Wolf and R. M. Philpot, *Mutation Res.*, 1980, **77**, 55.
[10] E. R. Nestmann, E. R. Stephen, D. J. Kowbel and A. Nasim, *Can. J. Genet. Cytol.*, 1982, **24**, 771.
[11] E. R. Nestmann and E. G.-H. Lee, *Mutation Res.*, 1983, **119**, 273.
[12] E. R. Nestmann and E. G.-H. Lee, *Mutation Res.*, 1985, **155**, 53.
[13] R. C. von Borstel, in *Short-Term Tests for Chemical Carcinogens*, H. F. Stich and R. H. C. San (eds.), Springer-Verlag, New York, 1981, p. 161.
[14] G. R. Douglas, E. R. Nestmann, J. L. Betts, J. C. Mueller, E. G.-H. Lee, H. F. Stich, R. H. C. San, R. P. Brouzes, A. L. Chmelauskas, H. D. Paavila and C. C. Walden, in *Water Chlorination: Environmental Impact and Health Effects*, R. L. Jolley, W. A. Brungs and R. B. Cumming (eds.), Ann Arbor Science, Ann Arbor, 1980, Vol. 3, p. 865.

[15] G. R. Douglas, E. R. Nestmann, O. P. Kamra, A. B. McKague, E. G.-H. Lee, J. A. Ellenton, R. Bell, D. Kowbel, V. Liu and J. Pooley, in *Short-Term Bioassays in the Analysis of Complex Environmental Mixtures,* M. D. Waters, S. S. Sandhu, J. Lewtas, L. Claxton, N. Chernoff and S. Nesnow (eds.), Plenum Press, New York, 1983, Vol. III, p. 431.

[16] O. P. Kamra, E. R. Nestmann, G. R. Douglas, D. J. Kowbel and T. R. Harrington, *Mutation Res.,* 1983, **118,** 269.

[17] G. R. Douglas, E. R. Nestmann, A. B. McKague, R. H. C. San, E. G.-H. Lee, V. W. Liu-Lee and D. J. Kowbel, in *Carcinogens and Mutagens,* Vol. IV, *The Workplace,* H. F. Stich, (ed.), CRC Press, Boca Raton, 1984, in the press.

[18] E. R. Nestmann, E. G.-H. Lee, T. I. Matula, G. R. Douglas and J. C. Mueller, *Mutation Res.,* 1980, **79,** 203.

[19] E. R. Nestmann, E. G.-H. Lee, J. C. Mueller and G. R. Douglas, *Environ. Mutagenesis,* 1979, **1,** 301.

[20] K. P. Kringstad, P. O. Ljungquist, F. de Sousa and L. M. Stromberg, *Environ. Sci. Technol.,* 1981, **15,** 562.

[21] V. F. Simmon, K. Kauhanen and R. G. Tardiff, in *Progress in Genetic Toxicology, Developments in Toxicology and Envrionmental Science,* D. Scott, B. A. Bridges and F. H. Sobels (eds.), Elsevier/North-Holland, Amsterdam, 1977, Vol. 2, p. 249.

[22] H. Brem, A. B. Stein and H. S. Rosenkranz, *Cancer Res.,* 1974, **34,** 2576.

[23] A. B. Swanson, D. D. Chambliss, J. C. Blomquist, E. C. Miller and J. A. Miller, *Mutation Res.,* 1979, **60,** 143.

[24] E. R. Nestmann, A. Nasim, R. H. Haynes and D. J. Kowbel, *Mutation Res.,* 1981, **89,** 229.

[25] F. J. de Serres and G. R. Hoffmann, in *Evaluation of Short-Term Tests for Carcinogens,* F. J. de Serres and J. Ashby (eds.), Elsevier/North-Holland, New York, 1981, p. 68.

[26] E. R. Nestmann, R. Otson, D. J. Kowbel, P. D. Bothwell and T. R. Harrington, *Environ. Mutagenesis,* 1984, **6,** 71.

[27] J. McCann, E. Choi, E. Yamasaki and B. N. Ames, *Proc. Natl. Acad. Sci. U.S.A.,* 1975, **72,** 5135.

[28] A. Bjørseth, G. E. Carlberg and M. Møller, *Sci. Total Environ.,* 1979, **11,** 197.

[29] V. F. Simmon, *J. Natl. Cancer Inst.,* 1979, **62,** 901.

[30] B. R. Holmbom, R. H. Voss, R. D. Mortimer and A. Wong, *Tappi,* 1981, **64,** 172.

[31] M. M. Shahin and R. C. von Borstel, *Mutation Res.,* 1977, **48,** 173.

[32] M. Cerna and H. Kypenova, *Mutation Res.,* 1977, **46,** 214.

[33] A. A. Potter, E. R. Nestmann and V. N. Iyer, *Mutation Res.,* 1984, **131,** 197.

9

In vitro mutagenicity testing of airborne particulates

G. M. ALINK and **J. J. van HOUDT,** Department of Toxicology, Agricultural University, De Dreijen 12, 6703 BC Wageningen, The Netherlands

The ambient air generally contains a very complex mixture of components, some of which are identified as mutagens and carcinogens. The main sources of such air pollutants are a variety of combustion processes.

Epidemiological studies in several countries have shown that the incidence of lung cancer is higher in urban than in rural areas [1, 2]. 'The urban factor' has been explained as due to different smoking habits, occupational exposures and general air pollution [3]. Though tobacco smoking is known to be the dominant causal factor, it is believed that air pollution may be responsible for 5-10% of all cancers of the respiratory system [2]. Although little is known about the role of air pollution in the induction of other types of cancers it cannot be excluded that some types of pollutants are contributing factors for tumours at other sites beside the respiratory system.

Air pollution components are present as gases and as particulate matter. The distribution of many compounds between these phases depends on chemical and physical factors. Although gases may substantially contribute to the mutagenic potential, this chapter will deal only with the particulates. Particulate matter in polluted air consists of a variable mixture of elemental carbon and inorganic oxides and salts on which organic compounds are adsorbed. For toxicological assessment, the model for particle retention proposed by the ICRP Task Group on Lung Dynamics [3] can be used to calculate the concentrations actually deposited in the major regions of the respiratory tract. Deposition in the respiratory tract depends on the size of the particles, which varies between 1 and 15 μm. The smallest particles (<3 μm) are retained in the pulmonary region. Of toxicological relevance are the facts that organic pollutants are usually most abundant on the smallest particles (<3 μm) [4] and that the smallest particles exhibit the highest mutagenicity [5]. Furthermore a rough approximation of the absorption efficiency indicates that about 70% of the material deposited in the pulmonary region will be absorbed, whereas the absorption in the nasopharyngeal and tracheobronchial tissues only amounts to 10%.

Risk assessment by testing individual compounds has serious drawbacks because of the large number of chemicals involved and the complexity of the mixture, in which various forms of interaction may occur such as synergism and antagonism.

Since only 30–40% of the organic compounds on airborne particles have been identified, the contribution of unindentified compounds to the toxicological risk may be significant. Therefore the assessment of the overall mutagenic or carcinogenic activity in air samples may provide a more realistic basis for the evaluation of the possible risks for man and environment, than testing individual compounds.

Short-term mutagenicity test systems with a certain predictive value for carcinogenicity have been developed, and can be used to assess mutagenicity of extracts of complex mixtures of particulate matter [6]. Using the very sensitive microbiological assay developed for mutagenicity testing by Ames [7], Pitts *et al.* [8] obtained preliminary data on mutagenic activity in the organic fraction of ambient airborne particulates. In the same year Tokiwa *et al.* [9] studied mutagenicity of particulate matter in several tester strains, and Dehnen *et al.* [10] studied different fractions of city smog extract.

In later years many other studies, mainly made with the microbial assay system, were undertaken [11, 12]. Mutagenic activity was observed in ambient particulate samples from industrial and urban areas [9, 10, 13-28]. Particulate matter from special sources such as automobile and diesel exhaust [29-33] and fly-ash from different combustion processes [34-36] have often been studied for mutagenicity.

However, few data on mutagenicity of airborne particulates are available for outdoor air samples taken at rural or background stations [8, 13, 16]. The first aim of our investigations was therefore to assess systematically the variation in mutagenesis of airborne particulates collected outdoors at non-industrial locations over a prolonged period of time. This study was performed in 1979 and 1980.

In view of the large fraction of time spent inside the home by people living in modern industrialized countries [37], exposure to mutagenic compounds in the air breathed might very well be determined predominantly by the concentration of mutagenic compounds in indoor air.

Several potential indoor sources of contaminants, such as cigarette smoking [38–41], volatilization of cooking products [42–45], wood-burning stoves [46] and burning joss sticks [47] have been shown to produce mutagenic compounds. Tobacco smoking seems to be one of the predominant sources.

Indoor suspended-particles concentrations in homes often exceed the outdoor concentrations when smokers are present in the home [48–51]. In the absence of smokers the indoor/outdoor ratio is usually less than 1, owing to filtration and deposition of particles entering the home from outside [52–55].

The mutagenic activity of airborne particles in homes has not been studied so far. Therefore the second purpose of our studies was an attempt to establish indoor/outdoor ratios for mutagenic activity of airborne particles. This study was done in 1982 and 1983.

9.1 MATERIALS AND METHODS

9.1.1 Sampling methodology

Sampling of outdoor particulate matter was done at two measuring sites: Wageningen and Terschelling (see Fig. 9.4, p. 133). The station at Terschelling, an island without industrial sites, was also used for background measurements of air pollution in the Netherlands. The sampling site near Wageningen was located at the meteorological station of the Agricultural University, which also functioned as a station of the Dutch Air Pollution Monitoring Network and a synoptic station of the Royal Dutch Meteorological Institute. Wageningen is a small rural town (approximately 30000 inhabitants) without important industrial sources of pollution. Both measuring sites are away from pollution sources such as traffic.

Samples of indoor particulate matter were collected in kitchens and living-rooms of 39 houses of employees of the Agricultural University of Wageningen.

For assessing the variation of outdoor mutagenicity, the standard high-volume technique was used. Over a period of 4 months at Terschelling and 12 months in Wageningen, samples were taken every 4th day. On the same days, SO_2 and O_3 concentrations, and meteorological parameters such as wind direction, wind speed, relative humidity and rainfall, were assessed for correlation with the mutagenicity data. For indoor sampling the high-volume sampler is not applicable owing to the high noise level and the large amount of air sampled, compared to the total air mass in kitchens and living-rooms. Therefore a sampler consisting of a centrifugal pump and a filter-holder with 1–3 glass-fibre filters was developed. Sampling was performed with the intake pointing down. For obtaining the data to be used in calculating the ratio of indoor/outdoor mutagenicity, outdoor particulate matter was not collected with a high-volume sampler but by following the 'indoor' procedure. Each week six samples were collected simultaneously. The occupants of the houses were asked to record the number of cigarettes, cigars and pipes smoked, cooking time, and some factors such as heating, ventilation and the use of solvents. Five categories of locations were sampled: heavy smokers' living-rooms (>100 cig/week; n = 6) moderate smokers' living-rooms (15 < cig/week <100; n = 5), non-smoker's living-rooms (<15 cig/week; n = 5), kitchens of non-smoking families (n = 23) and outdoors (n = 11).

The Gellman GF/A 20 × 24 cm filters used were equilibrated and weighed before and after sampling at room temperature and relative humidity of 45%. Differences in sampling methodology are summarized in Table 9.1.

9.1.2 Extraction

The filters with collected material were extracted with methanol for 8 hr in a Soxhlet apparatus. The methanol extracts were evaporated to dryness with a

Table 9.1 – Characteristics of high- and low-volume sampling

	High-volume sampling	Low-volume sampling
Sample velocity (m^3/hr)	100	10
Filter surface (cm^2)	480	1400
Linear velocity (m/sec)	0.5	0.025
Sampling time (hr)	24	168
Volume (m^3)	2400	1680
Storage temperature (°C)	4	–20

rotary vacuum evaporator. The residue was dissolved in DMSO. Extracts in DMSO were kept no longer than a few weeks before testing. The use of methanol was based on preliminary experiments and on literature data which also show methanol to be a suitable solvent.

9.1.3 Mutagenicity testing

9.1.3.1 Salmonella/microsome test

The extracts were tested in the Salmonella/microsome test as described by Ames *et al.* [56]. Because literature data and preliminary experiments indicated that strain TA98 in the most sensitive tester strain for particulate matter sampled in indoor and outdoor air, this strain was used in all experiments. Experiments were performed with and without the addition of Aroclor-induced liver homogenate. Two experiments were done per filter, each in triplicate. Extracts derived from different amounts of air were tested. In most cases single dose data were used for comparison between different samples. The outdoor mutagenicity is expressed as revertants per 30 m^3 of air, indoor mutagenicity is expressed as rev/m^3, both calculated from the slope of the linear part of the dose–effect curve. Ethyl methanesulphonate and dimethylnitrosamine were used as positive controls. For the indoor experiments benzo(*a*)pyrene and nitroquinoline oxide were used as positive controls. All data give the net number of revertants. Background mutagenicity varied between 16 and 52 (–S9) and between 28 and 62 (+S9) revertants per dish.

9.1.3.2 Sister-chromatid exchange test

Exponentially growing hamster cells (V79) were trypsinized. Amounts of 1 ml of a suspension containing 10^5 cells/ml were plated on object glasses placed in sterile 10-cm Petri dishes. After 2 hr, 9 ml of culture medium (Dulbecco's MEM) supplemented with 10% newborn calf serum (Gibco), penicillin (Gist-Brocades, 200 units/ml) and streptomycin (Gist-Brocades, 100 μg/ml) were added; 24 hr later, 0.1, 0.2 or 0.3 ml of dimethylsulphoxide, containing extracts from 20, 40 and 60 m^3 of air respectively, was added to the plates. The culture medium was replaced by a medium containing bromodeoxyuridine (BrdU, 10 μM) after 2 hr of exposure to the airborne particulate matter. The cell-cycle is 12 hr, so the exposure to BrdU was continued for 24 hr. Then the cells were treated according to the procedure of Perry and Wolff [57] with slight modifications [58] to make the sister-chromatid differential staining visible. Each sample was tested in 2 independent experiments. In each experiment cells were plated on 2 object glasses, and 50 metaphases were scored per experiment and per concentration. Six outdoor samples were tested: four samples for eastern and two for northern winds. The samples were taken in the summer.

9.2 RESULTS

9.2.1 Variation of outdoor mutagenicity

9.2.1.1 Background vs rural location

In Fig. 9.1 the number of revertants per 30 m^3 of air is given for the background location (Terschelling) and the rural location (Wageningen) during the months July–October. Although the difference between the geometric mean at Terschelling (24 rev/30 m^3) and Wageningen (104 rev/30 m^3) was significant ($t = -3.476$, $n = 22$, $P < 0.005$), the mutagenicity (log rev/30 m^3) at both locations was correlated ($r = 0.80$, $n = 22$, $P < 0.01$). On days when there was a higher mutagenicity the values at Terschelling were generally higher than those at Wageningen.

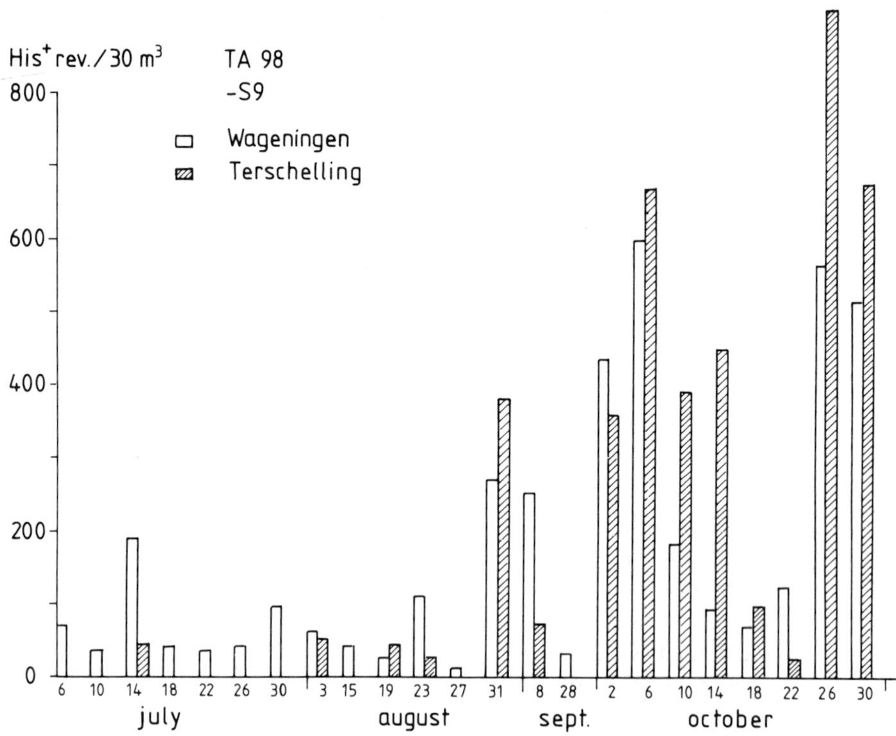

Fig. 9.1 – Mutagenic activity of extracts of particulate matter from 30 m^3 of air at Wageningen (rural location) and Terschelling (background location) in *S. typhimurium* TA98 (–S9) over a period of 4 months (July–October 1979).

Typical dose–effect curves for the mutagenicity at both locations on days with different wind directions are given in Fig. 9.2. On days with northern and western winds only a slight increase or none at all in the number of mutants was

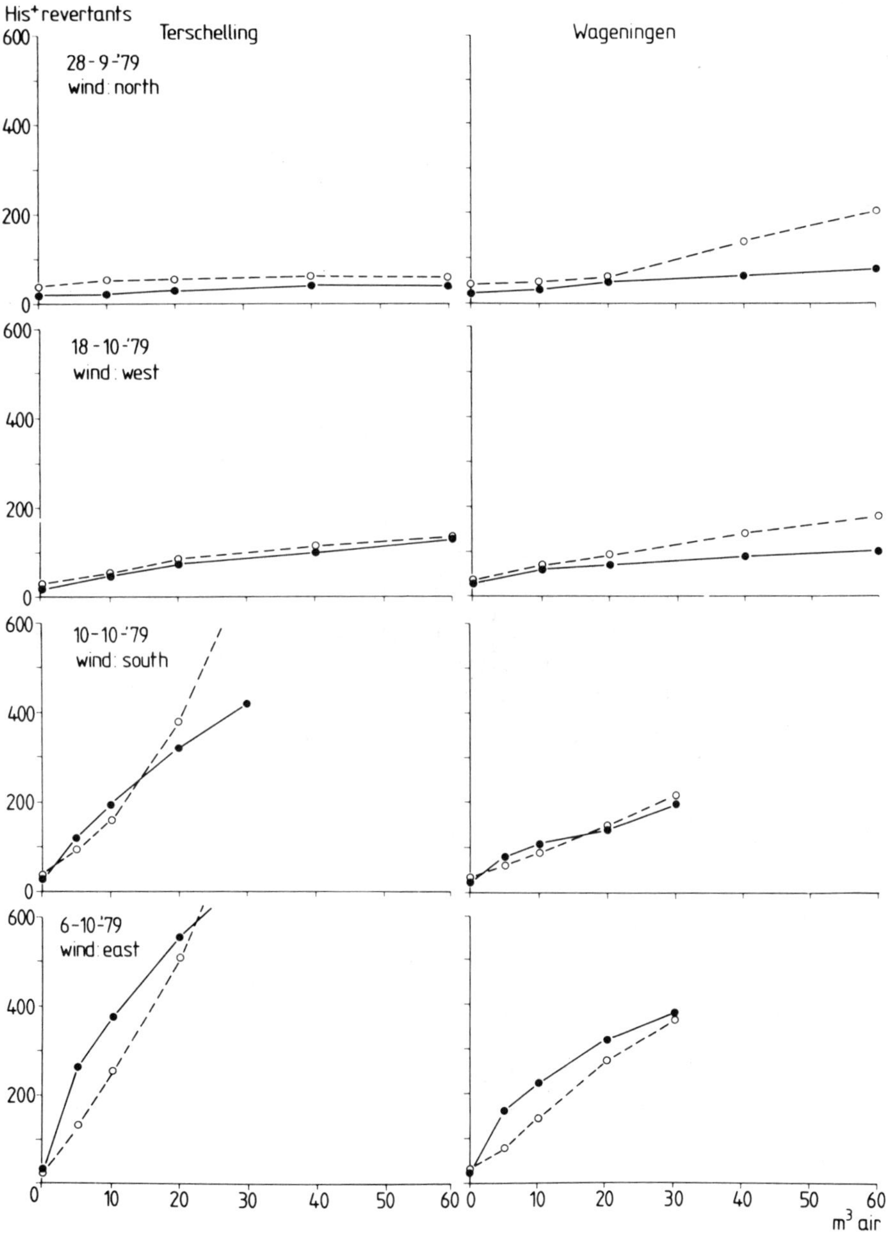

Fig. 9.2 – Mutagenic activity of extracts of airborne particulate matter at Wageningen and Terschelling in *S. typhimurium* TA98, with (○----○) and without (●——●) rat liver S9 fraction. Curves represent typical dose–effect relations for days with different wind directions.

found at the two locations. However, on days with a wind from the south or east the mutagenicity of the aerosols was generally much higher at both locations.

9.2.1.2 Mutagenicity over a one-year period

The mutagenicity of extracts of particulate matter obtained from 30 m³ of air at Wageningen on 78 days over a one-year period is given in Fig. 9.3. Although

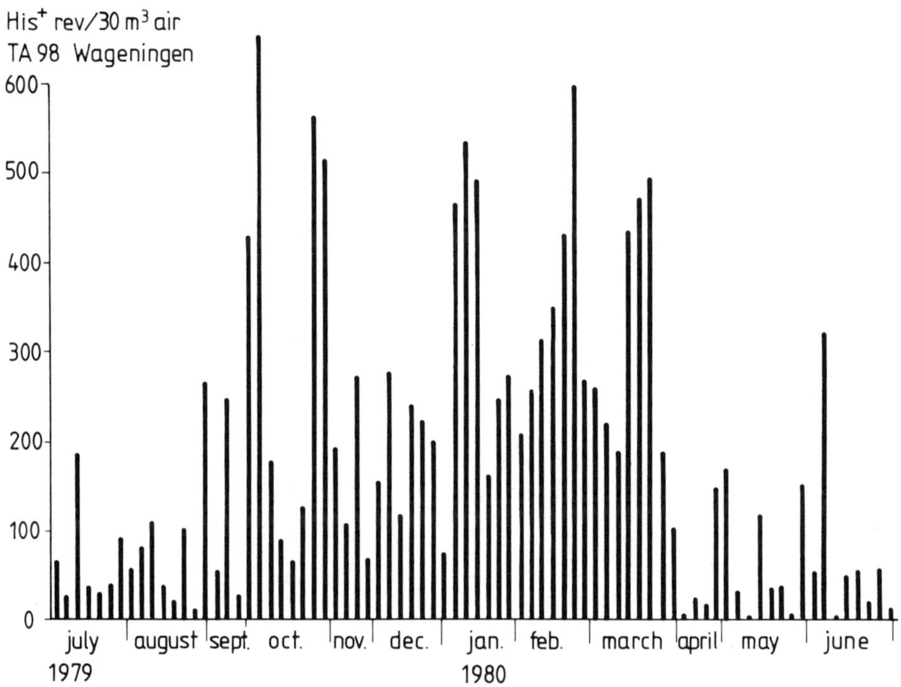

Fig. 9.3 – Mutagenicity of extracts of suspended matter from 30 m³ of air at Wageningen in *S. typhimurium* TA98 (–S9) over a 1-year period (July 1979–June 1980).

Table 9.2 – Mutagenicity of extracts of airborne particulate matter from 30 m³ of air at Wageningen on days with different wind directions (in sectors of 60°)

Sector no.	Sector	*n*	*m*	Range	S.D.
1	315-15°	5	44	0-153	62
2	15-75°	10	196	0-471	199
3	75-135°	7	396	119-597	196
4	135-195°	13	270	69-654	190
5	195-255°	19	143	1-316	95
6	155-315°	9	88	6-250	82

n = number of samples.

m = mean number of His⁺ revertants/30 m³ air-sample.

mutagenicity varied markedly from day to day, in winter it was significantly higher than in summer ($P < 0.01$). In Table 9.2 and Fig. 9.4, the mutagenicity is given for various wind directions, in segments of 60°. Samples were only attributed to a segment if the wind persisted for at least half the day in that segment. Samples collected on days with eastern and southern winds showed more mutagenic activity than samples obtained on other days.

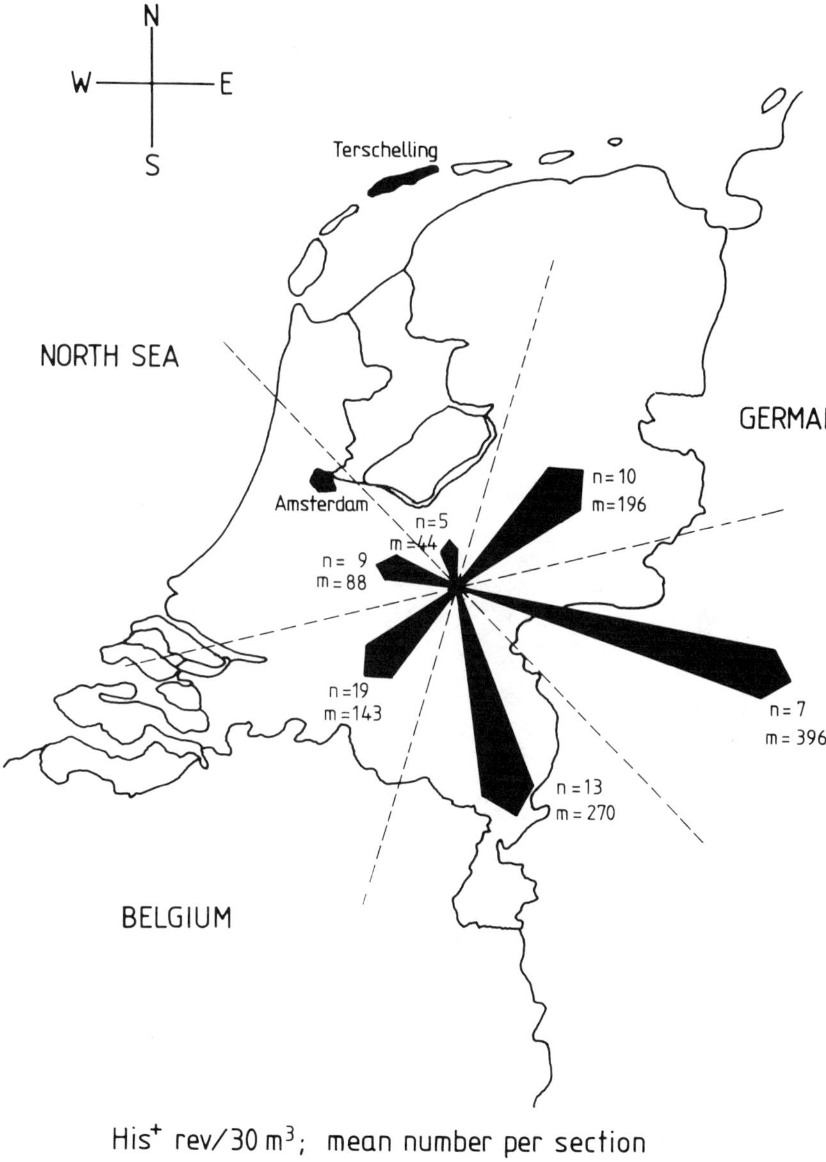

Fig. 9.4 – Mutagenicity of extracts of airborne particulate matter from 30 m^3 of air at Wageningen in *S. typhimurium* TA98 (–S9) on days with different wind direction. Arrows represent mean wind direction in sectors of 60°.

Other meterological parameters such as wind speed, relative humidity, and rainfall gave no correlation with mutagenicity. As shown by linear regression curves, the mutagenicity appeared to be significantly correlated with the SO_2 concentrations (average for 24 hr P <0.001), for the winter as well as for the

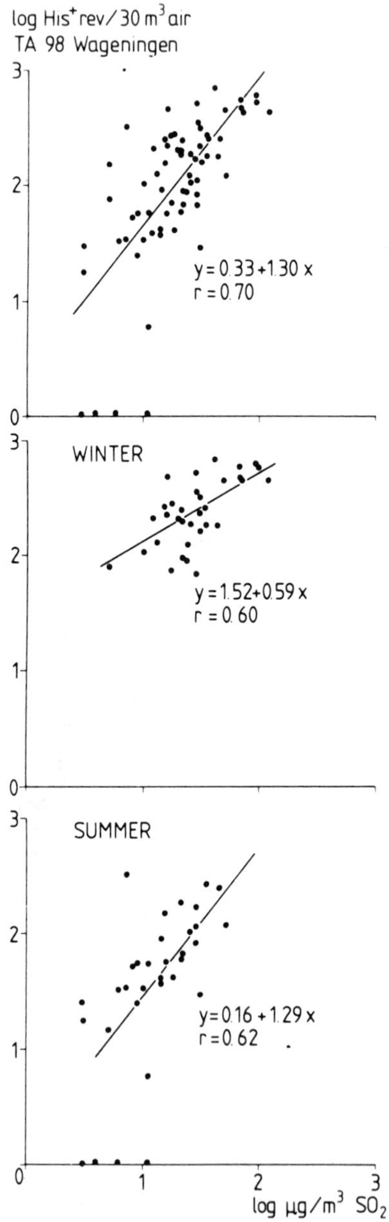

Fig. 9.5 – Linear regression curves, which give the correlation between the mutagenic activity of extracts of air particulate matter from 30 m³ of air, and the SO_2 concentration in winter, summer and a whole-year period.

summer period (Fig. 9.5). In Fig. 9.6 the corresponding curves are given for particle concentrations. In summer a lower correlation with mutagenicity was found than in winter.

As none of the sampling days had O_3 concentrations above 240 µg/m³ the

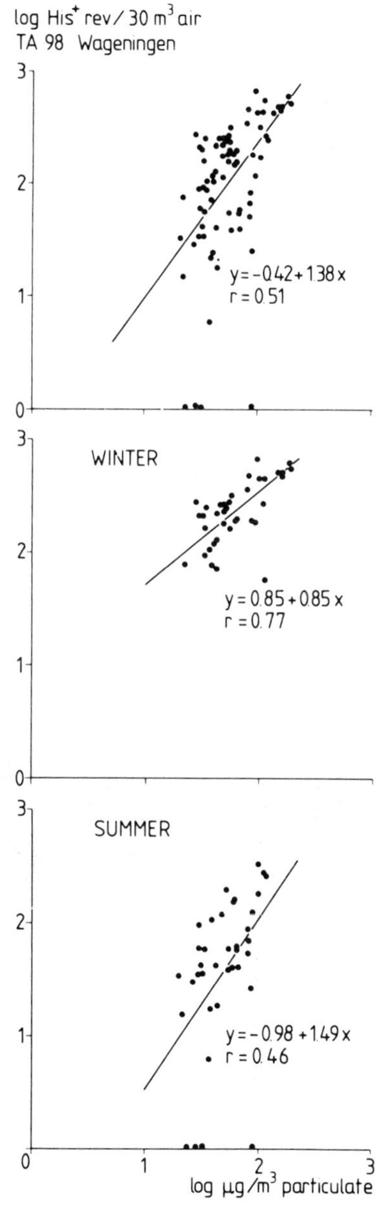

Fig. 9.6 – Linear regression curves, which give the correlation between the mutagenic activity of extracts of air particulate matter from 30 m³ of air, and the amount of particulate matter, in winter, summer and a whole-year period.

influence of photochemical air pollution could not be assessed. The effect of addition of metabolic enzymes to the assay plates is given in Fig. 9.7. Neither systematic change nor correlation with seasonal, meteorological or air-pollution parameters was found.

9.2.1.3 Induction of sister-chromatid exchange

The results for the four samples from eastern and two samples from northern winds in the SCE test and the Salmonella/microsome test are presented in Table 9.3. In Fig. 9.8, typical dose–effect curves are given for each wind direction. Samples that were positive in the Ames test were also positive in the SCE test, and samples that gave a weak or no mutagenic response in the bacterial test also gave only a small effect or none at all in the SCE test. The correlation coefficient between the dose-response relations of both tests varied between 0.95 and 0.99.

9.2.2 Indoor mutagenicity

9.2.2.1 Sampling data

The mean particle concentration for each sampling location and the results of the Salmonella/microsome assay are presented in Table 9.4. The results show

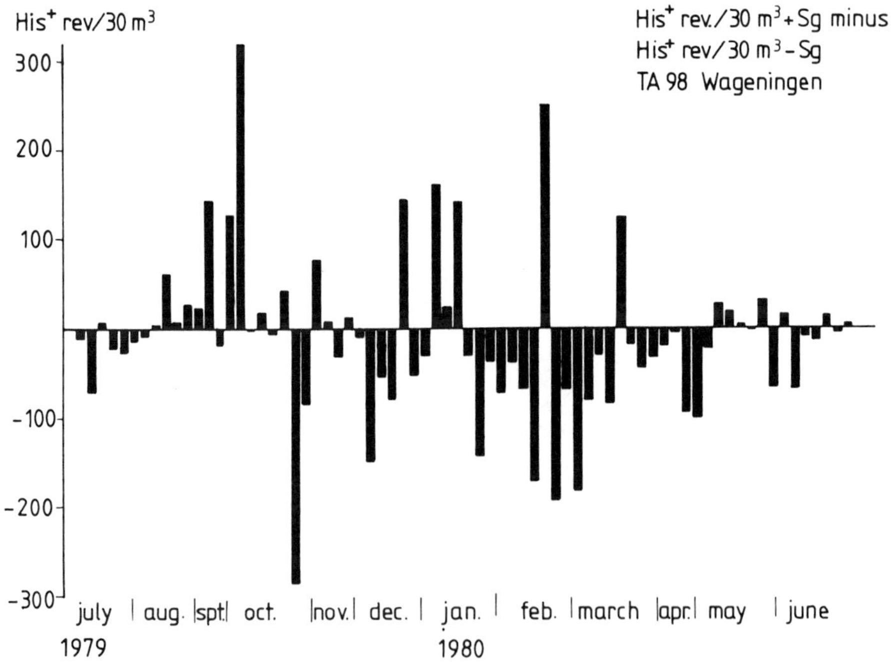

Fig. 9.7 – Effect of rat-liver S9 fraction on mutagenicity of extracts of particulate matter at Wageningen from 30 m³ of air. Data represent differences between revertants +S9 and −S9.

Table 9.3 – Mutagenicity of airborne particulates taken for eastern (E1–4) and northern (N1–2) winds in the SCE and Salmonella microsome test (mean ± standard error of mean, 50 replicates)

	SCEs/chromosome				His^+ revertants			
Sample volume (m^3)	0	20	40	60	0	20	40	60
E1	0.38 ± 0.03	0.77 ± 0.05	0.98 ± 0.04	1.15 ± 0.03	28 ± 2	151 ± 13	171 ± 4	195 ± 3
E2	0.40 ± 0.03	0.75 ± 0.04	0.98 ± 0.04	1.10 ± 0.04	27 ± 2	135 ± 5	183 ± 10	223 ± 11
E3	0.40 ± 0.03	0.71 ± 0.03	0.97 ± 0.04	1.12 ± 0.05	27 ± 3	182 ± 6	231 ± 19	263 ± 6
E4	0.41 ± 0.03	0.77 ± 0.05	1.11 ± 0.06	1.17 ± 0.06	26 ± 3	163 ± 7	230 ± 4	280 ± 14
N1	0.40 ± 0.02	0.54 ± 0.02	0.54 ± 0.03	0.60 ± 0.03	31 ± 3	62 ± 5	78 ± 6	74 ± 3
N2	0.43 ± 0.03	0.49 ± 0.02	0.52 ± 0.02	0.54 ± 0.03	34 ± 3	40 ± 7	43 ± 7	44 ± 2

Table 9.4 – Mean and range of aerosol concentration ($\mu g/m^3$) and mutagenic activity (rev/m^3) with and without addition of liver S9 for the various sampling locations. *Salmonella typhimurium* TA98 was used

Sampling location	number	concn. ($\mu g/m^3$)	rev/m^3–S9	rev/m^3 + S9
Outdoors	11	18 (6–48)	5 (0–14)	7 (2–23)
Kitchen	23	37 (6–83)	2 (0–23)	9 (2–30)
Living-room (cig/w <15)	5	40 (18–72)	2 (0–4)	10 (0–29)
Living-room (15 < cig/w <100)	5	50 (12–96)	5 (2–5)	17 (6–29)
Living-room (cig/w >100)	6	210 (42–354)	8 (0–22)	90 (14–202)

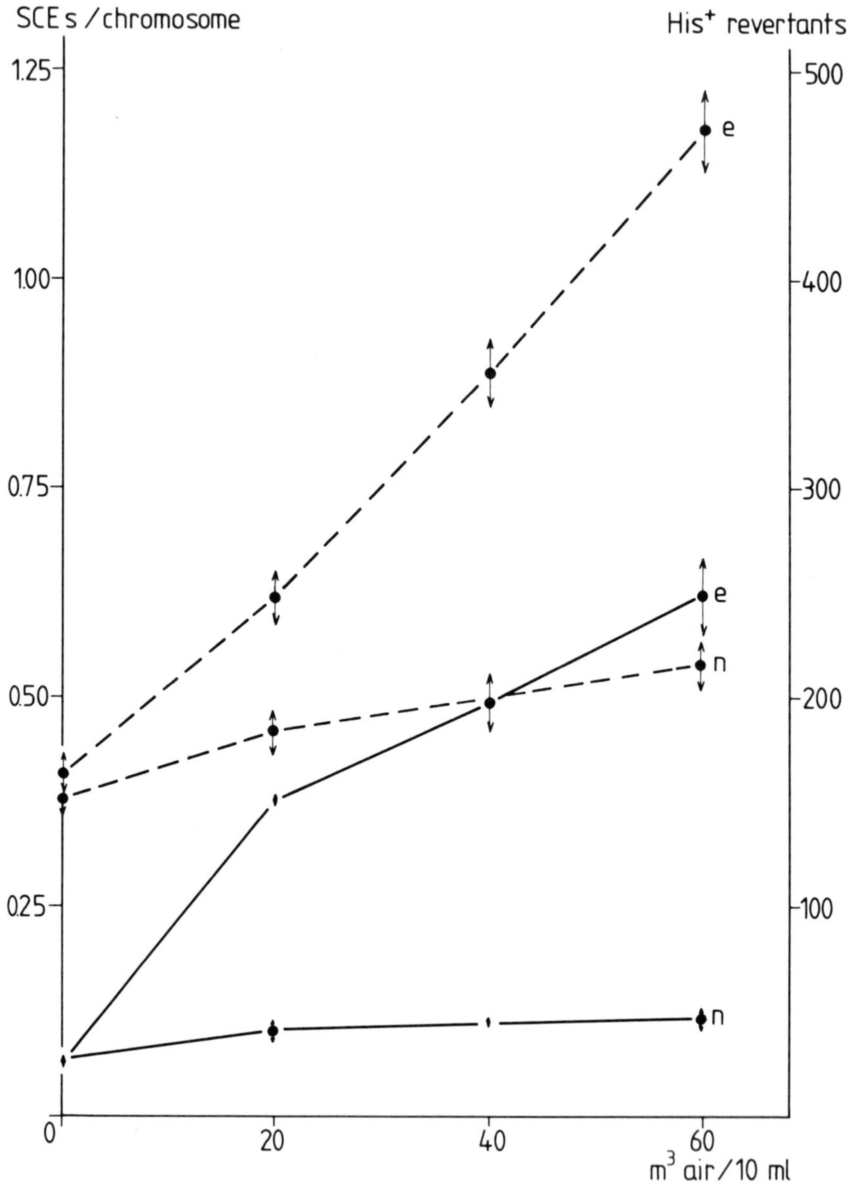

Fig. 9.8 – Mutagenicity of extracts of air particulate matter at Wageningen in *S. typhimurium* TA98 (–S9) (●———●) and Chinese hamster cells (V79 – SCE test) (○– – – –○) for eastern (e) and northern (n) winds. Curves represent typical dose–effect relations. Mean ± standard error of the mean.

that indoor concentrations of total suspended particles exceed outdoor concentrations and that cigarette smoking seems to be an important source. Concentrations in kitchens are comparable with those in non-smokers' living-rooms. Most living-room samples were collected in October–November 1982 whereas most kitchen samples were collected in January–February 1983. Airborne particulate matter, collected in kitchens, living-rooms and outdoors, is contaminated with compounds which are mutagenic in the Salmonella/microsome assay. Most outdoor samples show a detectable direct activity, which is slightly increased after addition of liver S9. Indoor samples show low or undetectable direct mutagenic activity. After addition of S9 a strong increase of the mutagenic activity was observed in all indoor samples except one. Consequently the ratio –S9/+S9 is much lower for indoor than for outdoor particles.

9.2.2.2 Effect of smoking

Figure 9.9 presents a typical example of the effect of smoking habits on indoor mutagenic activity. The dose–effect curves of the extracts simultaneously collected during one week are given. The figure clearly shows that in the outdoor

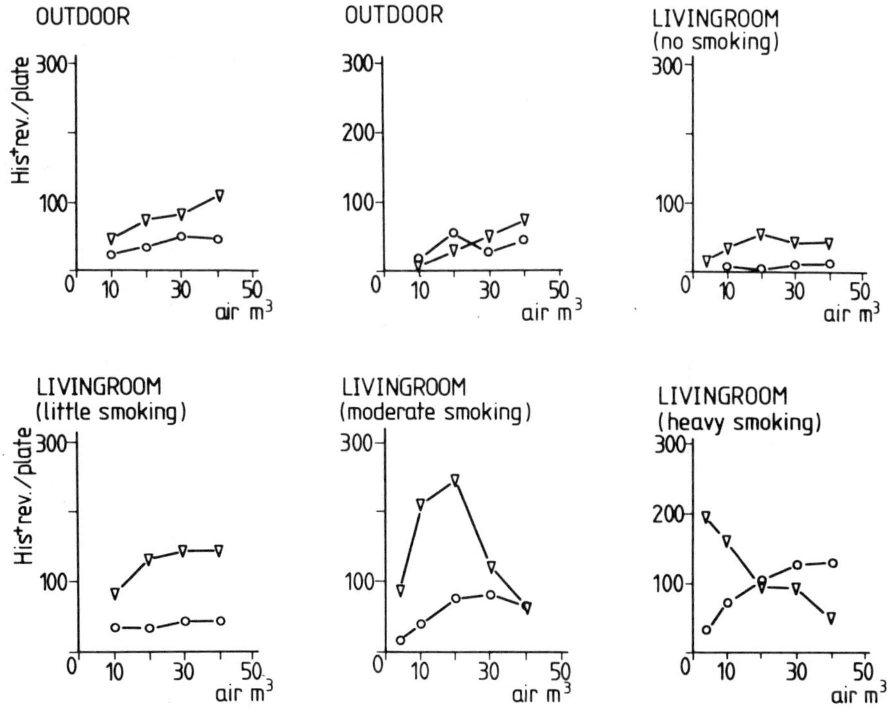

Fig. 9.9 – Mutagenic activity of extracts of airborne particulate matter, collected simultaneously in different types of living-rooms, compared to outdoor mutagenicity. During collection of samples in living-rooms, 0, 29, 90 and 160 cigarettes, respectively, were smoked. *Salmonella typhimurium* TA98 with (–△–) and without (–○–) liver S9 was used.

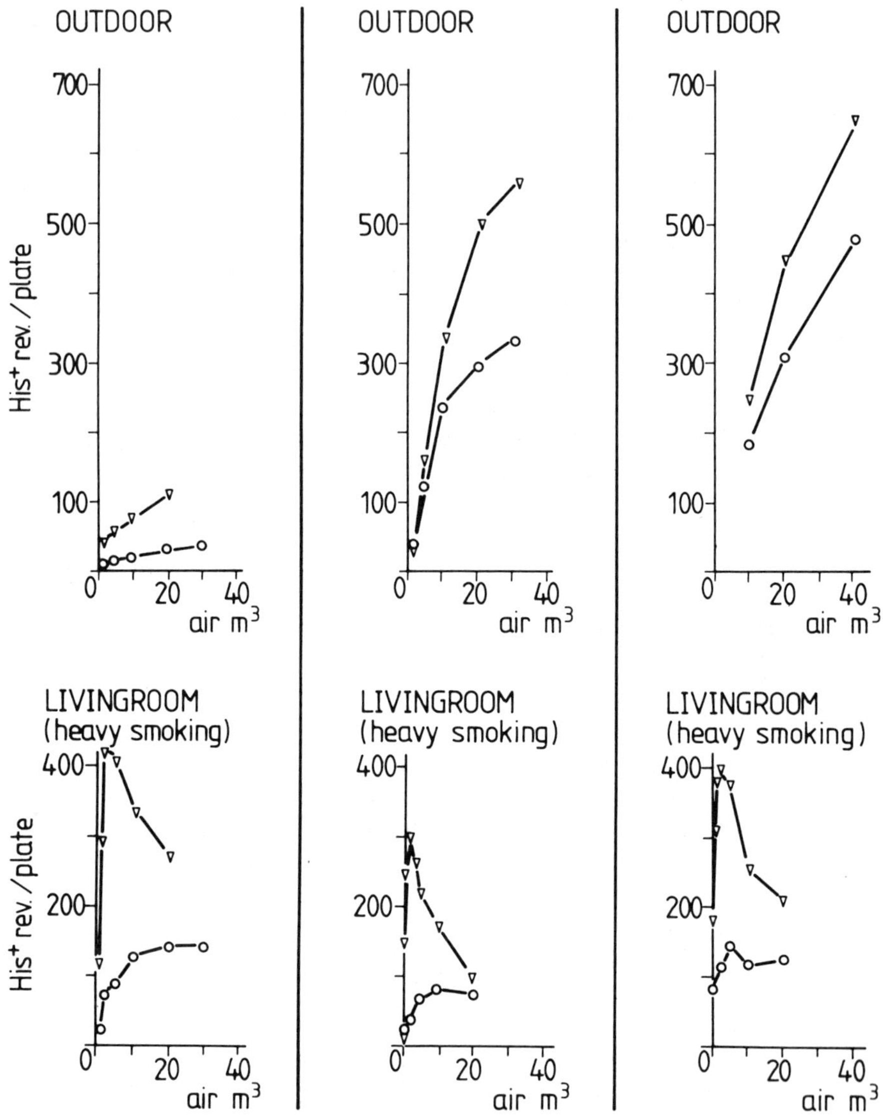

Fig. 9.10 – Mutagenic activity of airborne particulate matter, sampled in living-rooms of heavy smokers, compared with simultaneously sampled outdoor particles. *Salmonella typhimurium* strain TA98 with (–△–) and without (–○–) liver S9.

samples only a slight mutagenic activity is detectable. The indoor samples showed only a slight direct activity, but enzyme-mediated mutagenic effects for some samples were strong and related to the number of cigarettes smoked. The decrease in the number of revertants at higher doses after metabolic activation of samples collected in smokers' living-rooms may be caused by cytotoxic effects. The tested concentrations in the heavy smokers' living-room were all cytotoxic; the maximum of the curve was probably not found, but is at least 195 rev/plate, corresponding to a mutagenic activity of 49 rev/m^3. Some more examples of the influence of smoking are given in Fig. 9.10, which shows the results obtained in the living-rooms of three heavy smokers (>100 cig/week) compared to the appropriate outdoor samples. Notwithstanding the degree of outdoor indirect activity, in all cases indoor enzyme-mediated levels are much higher than the outdoors levels.

9.2.2.3 Kitchen effect

As mentioned, all kitchen samples were taken in homes of non-smokers. Figure 9.11 shows an example of the mutagenicity of two samples collected at different times in separate kitchens. The corresponding outdoor samples showed in one case a slight and in the other a strong mutagenic activity. The kitchen samples showed hardly any direct activity. However, after addition of liver S9 both kitchen samples had strongly increased mutagenic activity.

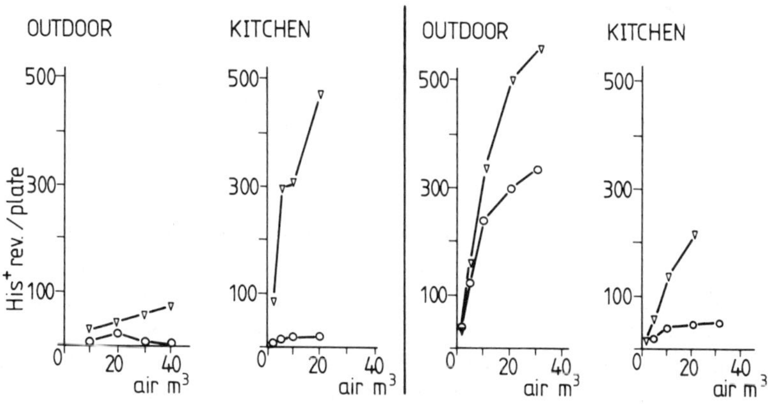

Fig. 9.11 – Mutagenicity of two kitchen samples, compared with simultaneously collected outdoor particles, in *Salmonella typhimurium* TA98 with (–△–) and without (–○–) liver S9.

To determine which sources are responsible for the mutagenic activity of particulate matter in kitchens, samples were taken for 4 weeks in kitchens of non-smoking families. Figure 9.12 gives a typical example of simultaneously collected samples. Pronounced cytotoxic effects, as found in smokers' living-rooms, were not observed. Two samples were collected in kitchens in which a ventilation hood was present. Possibly its presence caused the lower mutagenic activity in these samples.

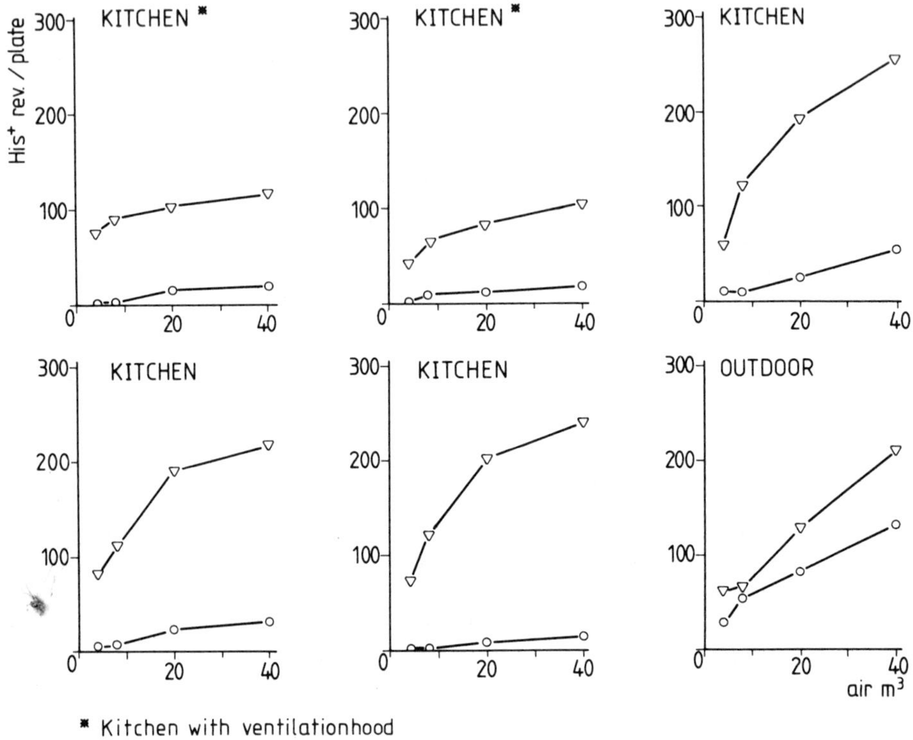

Fig. 9.12 – Mutagenicity in kitchens of non-smoking families and the corresponding outdoor sample, in *Salmonella typhimurium* TA98 with (–△–) and without (–○–) liver S9.

9.2.2.4 *Mutagenicity expressed as revertants/μg of particulate matter*

So far mutagenic activity has been expressed as revertants per cubic metre of air. Although this measure is most suitable in terms of personal exposure, it might be interesting to look at the mutagenic response per unit mass of suspended particulate matter. Both indoor particulate matter concentration and indoor mutagenic activity are higher than those outdoors, especially in smokers' homes. The question arises whether variations in mutagenicity at different locations may be due to differences in particulate matter concentration. Table 9.5 gives the number of revertants per μg of suspended particulate matter. It indicates that indoor direct-mutagenic activity unit per mass of suspended particles is lower than that outdoors, whereas the indirect mutagenicity per unit mass is comparable.

9.3 DISCUSSION

Much evidence exists on the mutagenicity of airborne particulates at urban and industrial locations. In recent years the number of publications on this subject

Table 9.5 – Mean number of revertants per 10 μg of particulate matter for the various locations. *Salmonella typhimurium* TA98 was used

Sampling location	–S9	+S9
Outdoors	2.1	4.1
Kitchen	0.5	2.4
Living-room (cig/w <15)	0.5	2.6
Living-room (15 < cig/w <100)	0.6	3.2
Living-room (cig/w >100)	0.4	5.5

has increased substantially [11, 12]. Only a few studies deal with background or rural areas [8, 13, 16]. In this study, using the Ames assay and the SCE test, we collected data showing that particulate matter at background and rural locations may also bear mutagenic compounds. This was previously found by Alfheim and Møller [13] and Dehnen *et al.* [16].

The correlation between mutagenicity of samples taken on the same day at different locations, a rural site and a background site, suggests that the mutagenic potential of suspended matter originates not from local sources but rather depends on a large-scale process. Our data provide some support for the suggestion of Alfheim and Møller [13] that mutagenicity at non-urban locations may be due to long-range transport of mutagenic particulate matter.

The long-range transport of air pollution in general in The Netherlands was already known. Van Egmond *et al.* [59] demonstrated the influence of industrial areas in Belgium and Germany. When mutagenicity is studied over a one-year period at a rural location the variation in mutagenicity between different days over the entire period is considerable. Our studies indicate that, in the winter period, mutagenicity is significantly higher than in the summer period. Probably this is due to seasonal differences in generating sources, such as combustion processes for space heating. Differences between summer and winter were also found at other locations, by Alfheim and Møller [13], Daisey *et al.* [15], Møller and Alfheim [20] and Dehnen *et al.* [16].

This study has shown that meteorological factors such as wind direction are of importance for the mutagenic burden. Commoner *et al.* [14] had previously made observations on the relationship between mutagenicity and wind direction. Alfheim and Møller [13] also indicated the significance of wind direction for the mutagenic potential.

This study shows that, at non-industrial locations, suspended matter may also contain direct-acting mutagens. This is in agreement with data published by others for urban and industrial areas [8–10, 14–17, 20, 23–25, 27]. The addition of metabolic enzymes leads to either an unchanged, a decreased or an enhanced number of revertants. These effects were correlated neither with a season, nor with a meteorological or air pollution factor. The unpredictable effect of liver microsomes on the mutagenic potential of extracts may be due to the changing composition of the extracts with substances that are directly mutagenic and others that require the presence of microsomes, whereas other directly mutagenic substances or non-mutagenic toxic substances may be activated when microsomes are present.

Tokiwa *et al.* [25] and Teranishi *et al.* [24] observed that different fractions prepared from the crude extract react in a different way on the addition of microsomes. Acid fractions and substituted polycylic aromatic hydrocarbons (PAHs) always display direct mutagenicity, whereas unsubstituted PAHs and the oxygenated fractions need S9. Wang *et al.* [30] found a decrease in mutagenicity of some automobile exhaust samples containing a high number of direct-acting mutagens, after addition of microsomes. The same decrease was found for many samples associated with combustion processes [34, 36].

Wei *et al.* [33] showed a possible non-enzymatic antimutagenic effect of S9 for diesel exhaust. Extracts of airborne particulates may display toxic effects [10, 60] which may be inhibited after addition of S9 [10].

Our studies show that airborne particles inside houses as well as outdoors may bear mutagenic compounds. Comparison of mutagenic activity between particulate matter from indoor and outdoor samples indicates that mutagenicity in indoor environments differs from that outdoors. The first difference is in the number of revertants per cubic metre of air. Outdoor samples show a direct mutagenic activity, which is clearly detectable in nearly all samples. Most indoor samples, however, show only low or undetectable direct activity, though in some cases they equal or exceed outdoor samples in direct activity. In contrast to outdoors samples, the increase after metabolic activation of indoor sample mutagenicity is very large; the highest indirect mutagenic response is found indoors, especially in smokers' homes. Even in the absence of smokers a relatively high mutagenic activity after metabolic activation is found, particularly in kitchens. In the first experiments, the 'outdoor' experiments, the strongest mutagenic activity of 78 extracts of particulate matter (daily average), sampled with a high-volume sampler over a year at the local meteorological station, was 25 rev/m^3 (median 5 rev/m^3) in the presence of S9 and 22 rev/m^3 (median 3 rev/m^3) without S9. The highest value indoors was 211 rev/m^3 with and 57 rev/m^3 without metabolic activation. The differences in results are not conclusive however, because of the differences in methodology (e.g. particle size) between high- and low-volume sampling. Outdoor extracts contain direct and indirect action mutagens, whereas indoor samples seem to exclusively contain mutagens which require metabolic activation. Löfroth *et al.* [61] found indications of a strong mutagenic activity after metabolic activation in particulate matter collected in office buildings. They also found in cigarette sidestream condensate about 15000 revertants per cigarette. No response was detected in the absence of S9. This is in good agreement with our results, which show an influence of cigarette smoke on indirect mutagenic activity of indoor-air particulate matter.

Besides extracts collected in smokers' living-rooms, kitchen extracts also show a very weak direct activity which increases after addition of liver S9. This mutagenic activity is produced by indoor sources. Some kitchen samples show a higher mutagenic activity than outdoor samples. It is known that extracts of volatilization products show mutagenic activity. A considerable amount of indirectly acting mutagens can be produced during cooking [43].

The mutagenicity per unit mass (Table 9.5) gives support to the conclusions mentioned above. Again it is obvious that indoor particles do not contain large numbers of directly acting mutagens, in contrast to outdoor samples. The mutagenicity per unit mass also indicates that the high level of enzyme-mediated activity in indoor samples is not attributable to the mutagenicity of the particulate matter itself, but to the high concentration of total suspended matter. In this study it was observed that indoor concentrations of particles exceed outdoor concentrations. This was previously found by several other authors [48, 51, 53]. Various publications indicate that indoor air is contaminated by particles of outdoor origin [52-55]. In our study, however, identification of outdoor particles as a source of indoor mutagenic activity is contradicted by the fact that outdoor samples show direct mutagenicity, whereas simultaneously sampled indoor particles do not show direct activity.

In these studies mutagenic activity was observed in complex environmental mixtures from indoors and outdoors. To assess whether or not extracts of airborne particulate matter represent a health risk for humans, the observed bacterial mutagenicity has to be evaluated by using *in vitro* test systems with mammalian cells. Furthermore it might be necessary to determine which chemical compounds or groups contribute to the mutagenic burden. Finally it may be stated that more attention has to be paid to the contribution of indoor sources of mutagenicity to the possible health impact of airborne particulates.

REFERENCES

[1] J. Higginson and O. M. Jensen, in *Air Pollution and Cancer in Man,* U. Mohr, D. Schmähl and L. Tomatis (eds.), IARC, Lyon, 1977, p. 169.
[2] R. Doll, *Environ. Health Perspect.*, 1978, **22,** 23.
[3] ICRP Task Group on Lung Dynamics, Committee II, *Health Phys.*, 1966, **12,** 173.
[4] R. C. Pierce and M. Katz, *Environ. Sci. Technol.*, 1976, **10,** 45.
[5] R. Talcott and W. Harger, *Mutation Res.*, 1980, **79,** 177.
[6] D. G. Harnden, *Health Perspect.*, 1978, **22,** 67.
[7] B. N. Ames, in *Chemical Mutagens, Principles and Methods for their Detection,* A. Hollaender (ed.), Plenum Press, New York, 1971, Vol. 1, p. 267.
[8] J. N. Pitts, D. Grosjean and T. M. Mischke, *Toxicol. Lett.*, 1977, **1,** 65.
[9] H. Tokiwa, K. Morita, A. Takeyoshi, K. Takahashi and Y. Ohnishi, *Mutation Res.*, 1977, **48,** 237.
[10] W. Dehnen, N. Pitz and R. Tomingas, *Cancer Lett.*, 1977, **4,** 5.
[11] C. E. Chrisp and G. L. Fisher, *Mutation Res.*, 1980, **76,** 143.
[12] T. J. Hughes, E. Pellizzari, L. Little, C. Sparacino and A. Kolber, *Mutation Res.*, 1980, **76,** 51.
[13] I. Alfheim and M. Møller, *Sci. Total Environ.*, 1979, **13,** 275.
[14] B. Commoner, P. Madyastha, A. Bronsdon and A. J. Vithayathil, *J. Toxicol. Environ. Health,* 1978, **4,** 59.
[15] J. M. Daisey, T. J. Kneip, I. Hawryluk and F. Mukai, *Environ. Sci. Technol.*, 1980, **14,** 1487.
[16] W. Dehnen, R. Tomingas and R. Pitz, *Zbl. Bakt. Hyg. I. Abt. Orig. B,* 1981, **172,** 351.
[17] D. Hoffmann, K. Norpoth, R. H. Wickramasinghe and G. Müller, *Zbl. Bakt. Hyg. I. Abt. Orig. B,* 1980, **171,** 388.
[18] H. Fukino, S. Mimura, K. Inoue and Y. Yamahe, *Mutation Res.*, 1982, **102,** 237.
[19] C. Ketseridis, M. Wullenweber and H. Rueden, *Zbl. Bakt. Hyg. I. Abt. Orig. B.*, 1982, **176,** 316.

[20] M. Møller and I. Alfheim, *Atmos. Environ.*, 1980, **14**, 83.
[21] M. Møller, I. Alfheim, S. Larssen and A. Mikalsen, *Environ. Sci. Technol.*, 1982, **16**, 221.
[22] J. N. Pitts, *Ecotoxicol. Environ. Safety*, 1978, **2**, 199.
[23] R. Talcott and E. Wei, *J. Natl. Cancer Inst.*, 1977, **58**, 449.
[24] K. Teranishi, K. Hamada and A. Watanabe, *Mutation Res.*, 1978, **56**, 273.
[25] H. Tokiwa, S. Kitamori, K. Takahashi and Y. Ohnishi, *Mutation Res.*, 1980, 77, 99.
[26] H. Tokiwa, S. Kitamori, K. Horikawa and R. Nakagawa, *Environ. Mut.*, 1983, **5**, 87.
[27] C. Y. Wang, M. S. Lee, C. M. King and P. O. Warner, *Chemosphere*, 1980, **9**, 83.
[28] W. K. de Raat, *Mutation Res.*, 1983, **116**, 47.
[29] J. Huisingh, R. Bradow, R. Jungers, L. Claxton, R. Zweidinger, S. Tejada, J. Bumgarner, F. Duffield, M. Waters, V. Simmon, C. Hare, C. Rodriquez and L. Snow, in *Application of Short-term Bioassays in the Fractionation and Analysis of Complex Environmental Mixtures*, EPA Publ. 600/9-78-027, 1978, p. 381.
[30] Yi Y. Wang, S. M. Rappaport, R. F. Sawyer, R. E. Talcott and E. T. Wei, *Cancer Lett.*, 1978, **5**, 39.
[31] Y. Ohnishi, K. Kachi, K. Sato, I. Tahara, H. Takeyoshi and H. Tokiwa, *Mutation Res.*, 1980, 77, 229.
[32] S. M. Rappaport, Yi Y. Wang and E. T. Wei, *Environ. Sci. Technol.*, 1980, **14**, 1505.
[33] E. T. Wei, Yi Y. Wang and S. M. Rappaport, *J. Air Pollution Control Assoc.*, 1980, **30**, 267.
[34] H. E. Kubitschek and L. Venta, *Environ. Mutagen.*, 1979, **1**, 79.
[35] H. E. Kubitschek and D. M. Williams, *Mutation Res.*, 1980, **77**, 287.
[36] G. Löfroth, *Chemosphere*, 1978, **10**, 791.
[37] D. J. Moschandreas, *Bull. N. Y. Acad. Med.*, 1981, **57**, 845.
[38] J. J. Hutton and C. Hackney, *Cancer Res.*, 1975, **35**, 2461.
[39] L. D. Kier, *Proc. Natl. Acad. Sci. U.S.A.*, 1974, **71**, 4159.
[40] T. Mizusaki, T. Takashima and K. Tomaru, *Mutation Res.*, 1977, **48**, 29.
[41] S. Mizusaki, H. Okamoto, A. Akiyama and Y. Fukuhara, *Mutation Res.*, 1977, **48**, 319.
[42] M. Nagao, M. Honda, Y. Seino, T. Yahagi and T. Sugimura, *Cancer Lett.*, 1977, **2**, 221.
[43] S. M. Rappaport, M. C. McCartney and E. T. Wei, *Cancer Let.*, 1979, **8**, 139.
[44] N. E. Spingarn and J. H. Weisburger, *Cancer Lett.*, 1979, **7**, 259.
[45] N. E. Spingarn, L. A. Slocum and J. H. Weisburger, *Cancer Lett.*, 1980, **9**, 7.
[46] L. Rudling, B. Ahling and G. Löfroth, *Int. Conf. Res. Solid Fuels*, 1–5 June 1981; *cf. Chem. Abstr.*, 1982, **97**, 97529t.
[47] S. Sato, R. Makino, Y. Takahashi, T. Sugimura and T. Miyazaki, *Mutation Res.*, 1980, 77, 31.
[48] J. D. Spengler, D. W. Dockerey, W. A. Turner, J. M. Wolfson and B. C. Ferris, *Atmos. Environ.*, 1981, **15**, 23.
[49] B. Brunekreef and J. S. M. Boleij, *Int. Arch. Occup. Envrion. Health*, 1982, **50**, 299.
[50] D. W. Dockerey and J. D. Spengler, *Atmos. Environ.*, 1981, **15**, 335.
[51] J. R. Girman, M. G. Apte, C. W. Traynor, J. R. Allen and C. D. Hollowell, *Environ. Int.*, 1982, **8**, 221.
[52] J. Alzona, B. L. Cohen, H. Rudolph, H. N. Jow and J. O. Frohlicher, *Atmos. Environ.*, 1979, **13**, 55.
[53] J. S. M. Boleij and B. Brunekreef, *Publ. Health Rev.*, 1982, **X**, No. 2, 169.
[54] A. F. Cohen and B. L. Cohen, *Atmos. Environ.*, 1980, **14**, 183.
[55] D. J. Moschandreas, J. W. Winchester, J. W. Nelson and R. M. Burton, *Atmos. Environ.*, 1979, **13**, 1413.
[56] B. N. Ames, J. McCann and E. Yamasaki, *Mutation Res.*, 1975, **31**, 347.
[57] P. Perry and S. Wolff, *Nature*, 1974, **261**, 156.
[58] W. M. F. Jongen, P. H. M. Lohman, M. J. Kottenhagen, G. M. Alink, F. Berends and J. H. Koeman, *Mutation Res.*, 1981, **81**, 203.
[59] N. D. Van Egmond, O. Tissing, D. Onderdelinden and C. Bartels, *Atmos. Environ.*, 1978, **12**, 2279.
[60] N. H. Seemayer and N. Manojlovic, *Toxicology*, 1980, **17**, 177.
[61] G. Löfroth, L. Nilson and I. Alfheim, in *Short Term-Bioassays in the Analysis of Complex Environmental Mixtures*, M. D. Waters, S. S. Sandhu, J. Lewtas, L. Claxton, D. Kowbel, V. Liu and J. Pooley (eds.), Plenum Press, New York, 1982, Vol. III, p. 515.

10

Genetic assays of maize for environment pollution control

M. J. PLEWA, Institute for Environmental Studies, University of Illinois, Urbana, Illinois 61801 U.S.A.

10.1 INTRODUCTION

Plant systems have been widely employed as mutagen indicators for over half a century. Higher plants possess a variety of well defined genetic end points, including the quantitative detection of chromosome damage and point mutations. Plant assays have several advantages as indicators of environmental mutagens and are the only higher eukaryotic systems currently employed as *in situ* monitors of polluted air [1-3], polluted water [4-6], agricultural chemicals [7-8], municipal sewage sludges [10, 11], and nuclear power plants [12]. Excellent reviews of the advantages of higher plants as monitors of environmental mutagens are available [4, 13].

An environmental mutagen, or genotoxin, is a physical or chemical agent released into the environment, that can alter the genome or its proper functioning. The presence of genotoxic agents in the environment constitutes a serious threat to the public health [14]. Depending on the ontogenetic stage of an organism, an environmental mutagen may exert teratogenic effects, affect the aging process, induce mutations in germinal cells [15], precipitate coronary disease [16] or cause mutations that lead to the neoplastic transformation of somatic cells [17, 18]. During the last 15 years a host of genetic assays have been developed to assess the mutagenic properties of chemicals [14]. A rational and prudent approach to limit the impact of environmental mutagens on the public health is simply to reduce the exposure of people to such agents.

The purpose of this paper is to describe a series of genetic assays based on the angiosperm *Zea mays* (maize) that may be suitable for use in environmental pollution control, and to present detailed case histories in which pesticides, municipal sewage sludges and chemical or physical mutagens are evaluated. The genetic end points that are described in this paper include reverse and forward mutation in microgametophytes (pollen grains), forward mutation in somatic cells, and chromosome aberrations in root-tip cells.

Table 10.1 – Pesticides Evaluated for Their Genotoxic Properties

Common Name	Chemical Name
INSECTICIDES	
Carbofuran	2,3-Dihydro-2,2-dimethylbenzofuran-7-yl methylcarbamate
Chlordane	1,2,4,5,6,7,8,8-Octachloro-2,3,3a,4,7,7a-hexahydro-4,7-methanoindene and related compounds
Chlorpyrifos	*O,O*-Diethyl *O*-(3,5,6-trichloro-2-pyridyl) phosphorothioate
Curacron	*O*-(4-Bromo-2-chlorophenyl) *O*-ethyl *S*-propyl phosphorothioate
Ethoprop	*O*-Ethyl *S,S*-dipropyl phosphorodithioate
Fonofos	*O*-Ethyl *S*-phenyl ethylphosphonodithioate
Heptachlor	1,4,5,6,7,8,8,-Heptachloro-3a,4,7,7a-tetrahydro-4,7-methanoindene and related compounds
Metham	Sodium methyldithiocarbamate
Phorate	*O,O*-Diethyl *S*-[(ethylthio)methyl] phosophorodithioate
Terbufos	*S*-{[(1,1-Dimethylethyl)thio] methyl} *O,O*-diethyl phosphorodithioate
HERBICIDES	
Alachlor	2-Chloro-2′,6′-diethyl-*N*-(methoxymethyl) acetanilide
Atrazine	2-Chloro-4(ethylamino)-6(isopropylamino)-*s*-triazine
Bifenox	Methyl 5-(2,4-dichlorophenoxy)-2-nitrobenzoate
Butylate	*S*-Ethyl-*N,N*-di-isobutylthiocarbamate
Cyanazine	2-[(4-Chloro-6-(ethylamino)-*s*-triazin-2-yl]-2-methylpropionitrile
Dicamba	2-Methoxy-3,6-dichlorobenzoic acid (3,6-dichloro-*o*-anisic acid)
Eradicane	*S*-Ethyl dipropylthiocarbamate
Metolachlor	2-Chloro-*N*-(2-ethyl-6-methylphenyl)-*N*-(2-methoxy-1-methylethyl)acetamide
Procyazine	2-{[(4-Chloro-6-(cyclopropylamino)-*s*-triazin-2-yl] amino}-2-methylpropionitrile
Propachlor	2-Chloro-*N*-isopropylacetanilide
SD50093	Formulation of Atrazine plus Cyanazine
Simazine	2-Chloro-4,6-bis(ethylamino)-*s*-triazine

10.2 *IN SITU* ANALYSIS OF PESTICIDES

A comprehensive analysis of the mutagenic properties of pesticides used in commercial corn production has been conducted [7-9]. This study encompassed the analysis of 21 pesticides and 12-combinations of herbicides. Three genetic assays were employed: reversed mutation in *Salmonella typhimurium*, gene conversion in *Saccharomyces cerevisiae* and reverse mutation in *Z. mays*. The use of these indicator organisms established a comprehensive data base that resolved a spectrum of genetic damage. The pesticides were evaluated by the microbial assays (a) directly, (b) after *in vitro* mammalian microsomal activation and (c) after *in vivo* plant activation. All the pesticides were also tested for their ability to induce mutations in maize under field (*in situ*) conditions. The research involving the *in situ* analysis of pesticides evaluated under normal, modern agricultural conditions is reviewed in this paper. The pesticides that were evaluated are presented in Table 10.1.

10.2.1 Maize *waxy* locus pollen assay

The *waxy (wx)* locus is located on the short arm of chromosome 9 of *Z. mays* at position 9-59. The *wx* locus controls the synthesis of the carbohydrate amylose. A pollen grain is a functional haploid, allowing the expression of either dominant or recessive alleles. The advantage of this assay is that the pollen grain is the unit of measurement, so large numbers can be analysed, providing a high degree of genetic resolution. A pollen grain carrying a *Wx* allele can synthesize amylose and this carbohydrate will form part of its starch, but a pollen grain carrying a recessive *wx* allele will not produce amylose. When an iodine solution is reacted with amylose a black complex results. Thus pollen grains that carry a *Wx* allele are stained black when exposed to an iodine solution, whereas pollen grains that carry a *wx* allele yield a tan colour. The starch type of a pollen grain is controlled by the genetic constitution of that pollen grain and not by the parental sporophyte. A genetic reversion (back mutation) of *wx* to *Wx* can be detected by counting the fraction of pollen grains from homozygous *wx* sporophytes, that are stained black by treatment with iodine solution (Fig. 10.1). Likewise, forward mutation can be measured by counting the fraction of tan coloured iodine-treated *wx*-carrying pollen grains obtained from plants that are homozygous for the *Wx* allele (Fig. 10.2).

10.2.2 *In situ* test plots

The *in situ* (in field) test plots for the evaluation of insecticides were constructed at the Illinois Natural History Survey. A series of rows was ploughed in an experimental field, and each subplot measured 2 x 0.5 m. Five kernels of homozygous *wx-C* inbred W22 maize were planted in each subplot. Control subplots were distributed within the field. A field-grade formulation of each insecticide was applied to its assigned subplot before emergence of the maize seedlings. The application rates expressed as kg/ha are listed in Table 10.2.

The *in situ* test plots for the evaluation of herbicides were constructed at the South Farms of the Department of Agronomy at the University of Illinois.

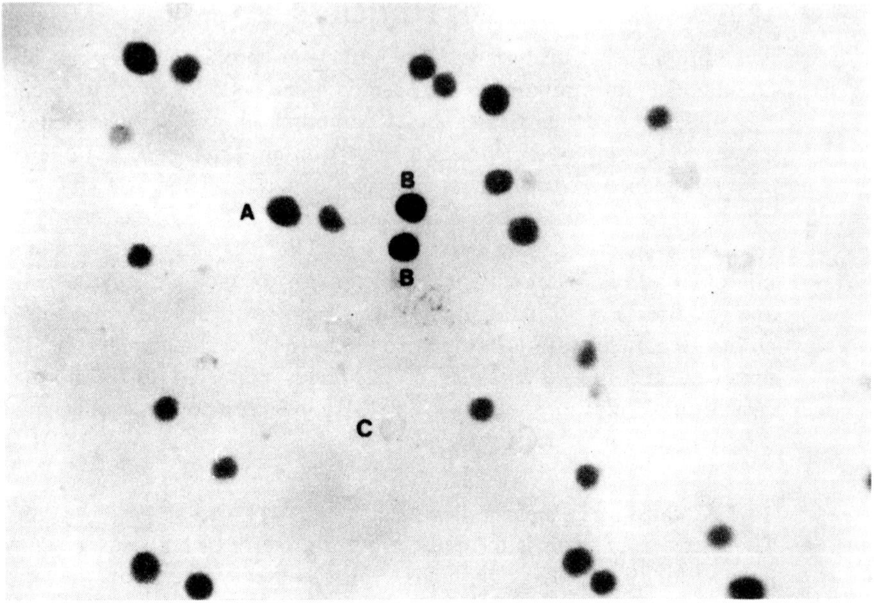

Fig. 10.1 – Reverse mutation at the *wx* locus in maize pollen grains: (A) a pollen grain carrying the recessive *wx-C* allele; (B) a reverse mutant pollen grain; (c) an aborted pollen grain. The pollen grains were from a plant that was *wx-C/wx-C*.

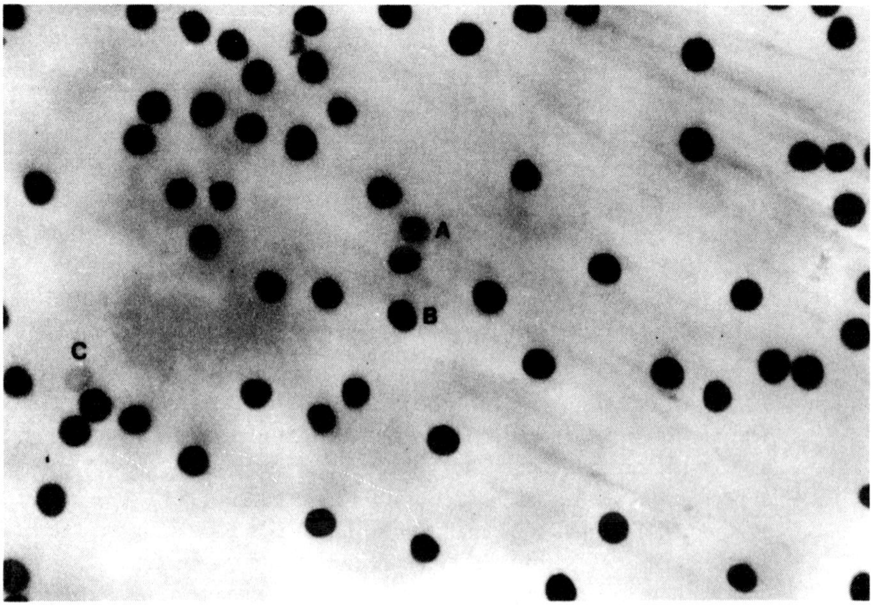

Fig. 10.2 – Forward mutation at the *wx* locus in maize pollen grains: (A) a forward mutant pollen grain; (B) a pollen grain carrying the dominant *Wx* allele; (C) an aborted pollen grain. The pollen grains were from a plant that was *Wx/Wx*.

Table 10.2 – Application Rates of Pesticides Evaluated

Pesticide	Application Rate (kg/ha)
PLOT 1	
Carbofuran	2.24
Chlorodane	2.24
Chlorpyrifos	2.24
Curacron	2.24
Ethoprop	2.24
Fonofos	2.24
Heptachlor	1.12
Metham	2.24
Phorate	2.24
Terbufos	2.24
PLOT 2	
Bifenox	2.24
Cyanazine	3.58
Procyazine	3.58
SD50093	4.48
PLOT 3	
Alachlor	6.00
Atrazine	3.84
Cyanazine	4.80
Eradicane	7.20
Metolachlor	8.40
Simazine	3.84
SD50093	4.80

Each plot was approximately 10 × 3 m and consisted of three parallel 10-m long rows. The outer two rows were planted with a commercial hybrid maize variety, and the centre row was planted with inbred W22, *wx-C/wx-C* kernels. A field-grade formulation of each herbicide was applied to its assigned subplot before emergence of the maize seedlings. Control subplots were distributed within the test plot. The application rates are listed in Table 10.2.

10.2.3 Collection and storage of tassels

The plants were allowed to grow until early anthesis and each tassel from the *wx-C/wx-C* plants was then harvested and immediately labelled. The tassels were dehydrated by storage in a tank of 70% ethanol for 2 days. The tassels were then stored in labelled 1-litre jars filled with 70% ethanol and the pollen grains were analysed at a later date.

10.2.4 Preparation of pollen slides

Each tassel was removed from its storage jar and agitated in clean ethanol, then approximately 15 unopened florets were removed. The anthers were dissected from these florets and placed in the stainless-steel cup of a VirTis microhomogenizer. The cup contained 0.6 ml of a gelatin–iodine stain prepared by mixing equal volumes of a stock iodine solution and a stock gelatin solution, and adding a drop of Tween-80. The stock iodine solution consisted of 25 ml of water, 500 mg of potassium iodide and 95 mg of iodine. The stock gelatin solution was prepared by mixing 25 ml of heated water and 1.5 g of bacteriological gelatin. Fresh stain was prepared daily. The anthers were minced with scissors and homogenized for 30 sec. The homogenate was strained through cheesecloth onto the surface of a large microscope slide and a coverslip was placed on the pollen suspension. After the suspension had solidified the slide was examined under a dissecting microscope at 40× magnification. The black pollen grains were counted and the total number of pollen grains on the slide was estimated by counting the number of pollen grains within 20 randomly-chosen 1-mm^2 areas and multiplying this value by a factor appropriate for the area of the coverslip. When a number of slides had been analysed for each plant, the frequency of revertant pollen grains was calculated by dividing the total number of *Wx* pollen grains by the estimated total number of pollen grains.

The frequency of mutant pollen grains for each group of plants from each subplot was compared with the frequency of the mutant pollen grains from the appropriate control subplots. The control and evaluation data were statistically analysed [19] and the data for the pesticides are presented in Figs. 10.3–10.5.

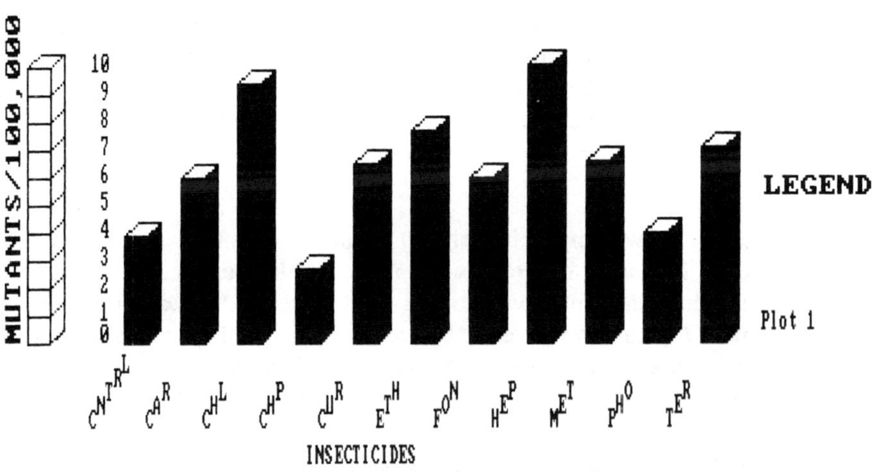

Fig. 10.3 – Mutation induction in maize plants by insectides, Plot 1. The bars indicate the frequency of revertant *Wx* pollen grains induced in *wx-C/wx-C* plants under *in situ* test conditions. The abbreviations for the pesticides are: CNTRL = Control, CAR = Carbofuran, CHL = Chlordane, CHP = Chlorpyrifos, CUR = Curacron, ETH = Ethoprop, FON = Fonofos, HEP = Heptachlor, MET = Metham, PHO = Phorate, TER = Terbufos.

An example of the data analysis for the control and *s*-triazine herbicides is presented in Table 10.3. In the tests in Plot 1, a significant increase in the frequency of mutant pollen grains was induced by chlordane, ethoprop and heptachlor [8] (Fig. 10.3). Of the herbicides evaluated in Plot 2, cyanazine and SD50093 exhibited mutagenic activity (Fig. 10.4). Of the herbicides evaluated in Plot 3, atrazine, simazine, cyanazine and SD50093 exhibited mutagenic activity [9] (Fig. 10.5).

Table 10.3 – Analysis of *s*-Triazine Herbicide Data

Treatment Groups	No. of Pollen Grains Analysed ($\times 10^5$)	Mutation Rate Among Gametophytes* ($\times 10^{-5}$)	Statistical Significance†
Control	23.89	4.24	NA
Atrazine	9.50	8.53	$P \leqslant 0.005$
Cyanazine	9.42	14.76	$P \leqslant 0.001$
Simazine	9.85	10.86	$P \leqslant 0.003$
SD50093	11.45	16.25	$P \leqslant 0.001$

*The estimation of the mutation rate among gametophytes is based on either the weighted or unweighted average mutation rate. The best estimator is the average with the least variance [20].
†Analysis of variance, $F_{4,35} = 6.93$. The probability indicates the level of significance between a control and treatment group.

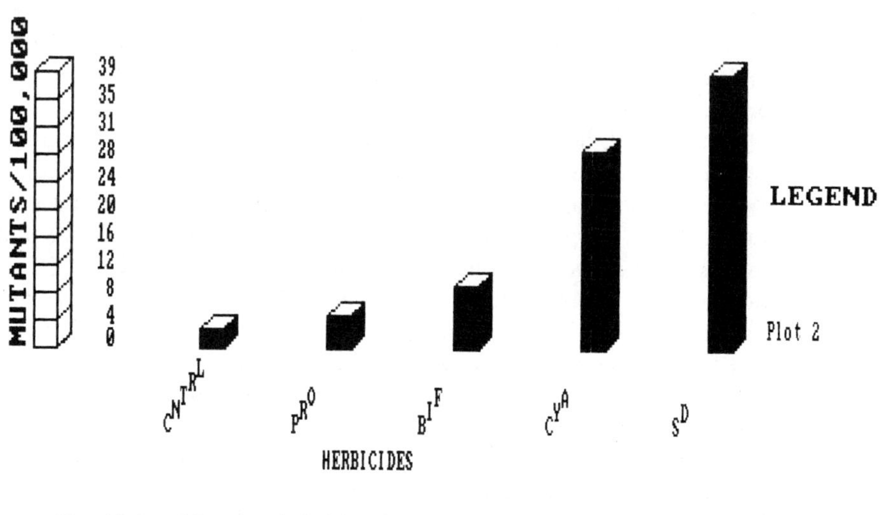

Fig. 10.4 – Mutation induction in maize plants by herbicides, Plot 2. The bars indicate the frequency of revertant *Wx* pollen grains induced in *wx-C*/*wx-C* plants under *in situ* test conditions. The abbreviations for the pesticides are: CNTRL = Control, PRO = Procyazine, BIF = Bifenox, CYA = Cyanazine, SD = SD50093.

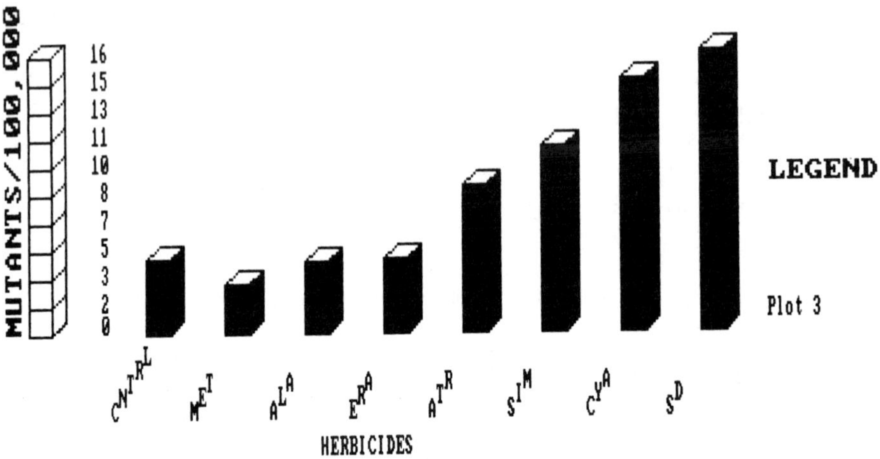

Fig. 10.5 – Mutation induction in maize plants by herbicides, Plot 3. The bars indicate the frequency of revertant *Wx* pollen grains induced in *wx-C/wx-C* plants under *in situ* test conditions. The abbreviations for the pesticides are: CNTRL = Control, MET = Metolachlor, ALA = Alachlor, ERA = Eradicane, ATR = Atrazine, SIM = Simazine, CYA = Cyanazine, SD = SD50093.

10.3 *IN SITU* ANALYSIS OF MUNICIPAL SEWAGE SLUDGE APPLIED TO SOIL

A problem facing the United States is the prudent disposal of the sludge resulting from the treatment of municipal sewage. In the past decade there has been a substantial increase in the number and efficiency of sewage treatment plants. An attractive disposal method is the application of this organic matter to agricultural land or its use in the reclamation of land stripped of its organic matter by surface mining.

There may, however, be problems associated with sludge disposal by land application. An *in situ* evaluation of the mutagenic potential of Chicago municipal sewage sludge applied to agricultural soils was conducted between May and August 1979 [11]. A test plot was constructed on the NW900 plots at the University of Illinois Agronomy Research Center near Elwood, IL. The plots had not been treated with pesticides since 1971 or sludge since 1974. *Z. mays* kernels of inbred M14, homozygous for either the *wx-C* or *wx-90* allele, were used for the study. One row of each allele type was planted in a control subplot and in each of three treatment subplots. Over the growing season, the maximum-treatment subplot received a total dosage of liquid sludge equivalent to 21.0 tons of dry material per hectare. The other treatment subplots received one-half and one-quarter of this amount. The control subplot did not receive any sludge, and adequate nutrients were supplied to the plants as chemical fertilizer. No pesticides were applied to the test plots during the experiment. The plants were allowed to grow until early anthesis and each tassel was individually labelled, harvested and stored in 70% ethanol for subsequent analysis. Pollen slides were prepared and analysed. The percentage of aborted pollen grains was also determined for each plant type.

For the *wx-C* allele, over 5.1×10^6 pollen grains were analysed. The fractions of mutant pollen grains did not increase with increase in the amount of sewage sludge added to the soil. However, a direct relationship was observed between the amount of applied sludge and the frequency of pollen abortion. The percentage of pollen abortion ranged from 19.4% for the control to 49.9% for the maximum-sludge subplot.

The *wx-90* allele responded differently to the sludge. For *wx-90* an increased frequency of revertants was observed with increased amount of sludge (Fig. 10.6).

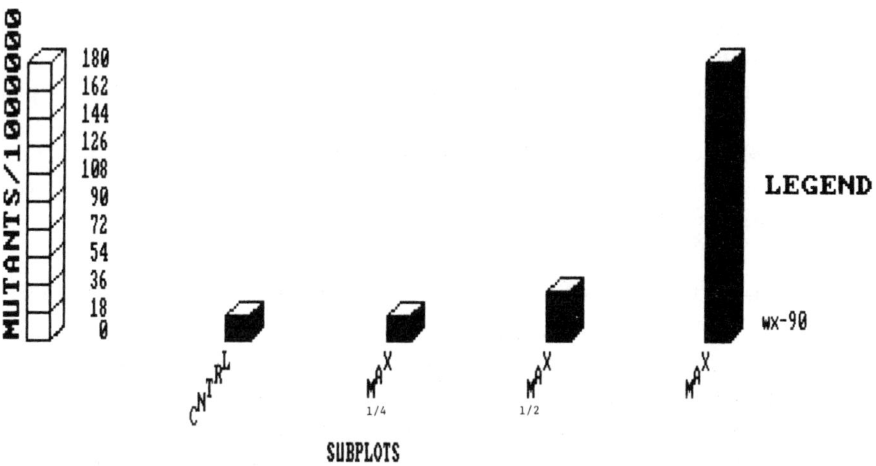

Fig. 10.6 – Mutation induction in maize plants by the treatment of agricultural soil with Chicago municipal sewage sludge. The bars indicate the frequency of revertant *Wx* pollen grains induced in *wx-90/wx-90* plants under *in situ* test conditions. The abbreviations for the treatment groups are: CNTRL = Control, 1/4 MAX = 5 tons dry of dry sludge per ha, 1/2 MAX = 10.5 tons/ha, MAX = 21 tons/ha.

In the control subplot, the frequency of revertant pollen grains was 1.6×10^{-6}. There was no increase in the frequency of revertant pollen grains in plants grown on the subplots treated with a quarter of the maximum sludge load. Plants grown in the subplots treated with half maximum sludge load exhibited a reversion frequency twice that for plants grown on the control subplot. There was a tenfold increase in the frequency, for plants grown in the maximum-treatment subplot. Thus the sludge-treated soil exhibited concentration-dependent increases in the induction of point mutation in maize germinal cells. The general conclusions of these *in situ* experiments were in agreement with those from concurrent analysis with other genetic indicator organisms [11].

10.4 RECOMMENDATIONS FOR *IN SITU* ANALYSES

The experiments above clearly demonstrate the utility of the *in situ* maize *wx* locus pollen assay for evaluating the environment for mutagenic contaminants. The most difficult component of the experimental designs is choosing suitable control subplots. When agents are applied to a series of subplots, non-treated subplots may serve as acceptable controls, but if interest lies in assaying an area suspected of contamination, it is necessary to find a satisfactory concurrent control in an uncontaminated area. One solution is to use greenhouses or plant-growth chambers. Although *in situ* environmental monitoring may become an attractive method for identifying the presence of genotoxic contaminants, several basic requirements must be fulfilled to ensure adquate experimental designs. These requirements include: (1) a knowledge of the history of contamination in the area to be assayed, (2) the construction of test plots with proper and concurrent controls, (3) the use of well defined assays with specific and calibrated genetic end points, (4) the use of appropriate numbers of indicator organisms to provide a sufficiently large data-base, (5) trained personnel to maintain the test plots and to label and harvest the plants at the correct time, (6) laboratory personnel to conduct competent analysis and proper interpretation of the observed genetic events, and (7) the use of appropriate statistical analysis in the evaluation of the data.

10.5 DEVELOPMENT OF A RAPID ENVIRONMENTAL MONITOR

10.5.1 Calibration experiments

The development of a rapid yet sensitive maize monitor for environmental mutagens, based on the inbred Early-Early Synthetic, is continuing. A discussion of the use of Early-Early Synthetic maize has been published [7]. Since this variety matures rapidly (from kernel to tassel emergence in less than 4 weeks) and is of short stature (approximately 50 cm in height) it can be used under *in situ* conditions as well as in the laboratory. The genetic end points that should be available for scoring include mutation induction at the *wx* locus in pollen grains, forward mutation at the *yg2* locus in leaves, and the induction of micronuclei in primary root-tip cells.

Forward mutation at the *wx* locus in pollen grains has been assayed after chronic exposure to chemical mutagens administered in the soil. The chemical mutagens used were ethyl methanesulphonate (EMS) and maleic hydrazide (MH). The use of these agents under controlled laboratory conditions gave a calibration of the forward mutation test of the *wx* locus assay. The experimental design included 20 plants, with 5 plants per treatment group. A concurrent control of 5 plants was included with each experiment. For each plant, one kernel was planted in a standard soil mixture of loam, peat-moss and sand (4:2:1 v/v/v) in a 10-cm diameter plastic pot. Each pot was enveloped in a plastic bag, and each five-pot treatment group was placed in a plastic tub. The tubs were placed in a plant-growth chamber adjusted to a photoperiod of 14 hr and an illumination of 300 $\mu E \cdot m^{-2} \cdot sec^{-1}$ photosynthetically relevant radiation (μE = microEinstein) with day and evening temperatures regulated at 25 and 20°C, respectively. All the plants were given identical amounts of water and fertilizer. The only variable was the concentration of mutagen administered to the soil. The chemical mutagen, EMS or MH, was applied to the soil in the pots as an aqueous solution of known concentration prepared immediately before application; 50-ml amounts were poured onto the soil of each pot three times a week. The control plants received water only, in a similar regime. When the tassels had emerged, treatment was stopped and the plants received water only. Since the amounts and concentration of each mutagen solution were recorded, the total amount of mutagen to which each plant was exposed through the soil could be calculated. After the tassels reached anthesis they were harvested and the pollen grains were scored for forward mutants and pollen abortions as described above (Fig. 10.2).

The results from the EMS experiments are presented in Fig. 10.7. The amount of EMS to which the plants were exposed ranged from 1.1 to 6.8 μmoles per plant. The frequency of mutant pollen grains in every treatment group was significantly different ($P \leqslant 0.001$) from that for the control group. The frequency of mutant pollen grains ranged from 5.11×10^{-5} for the control to 2.05×10^{-4} for the highest-treatment group. The concentration-response curve was linear ($r^2 = 0.93$) for the induction of forward mutant pollen grains. However, there was no correlation between the concentration of EMS and the percentage of aborted pollen grains ($r = 0.22$). Thus, EMS is a potent mutagen to Early-Early Synthetic maize when chronically administered in the soil. These data clearly indicate the sensitivity of this assay to environmental mutagens.

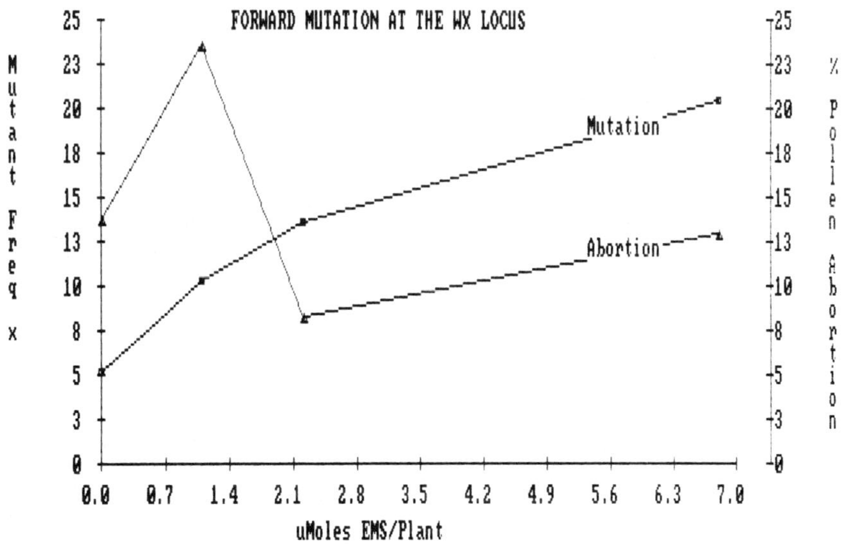

Fig. 10.7 – Concentration-response curve for chronic exposure of Early–Early Synthetic maize to ethyl methanesulphonate. Mutation represents the frequency of forward *wx* mutant pollen grains in plants that were *Wx/Wx*. Abortion represents the percentage of pollen abortions.

The results from the MH experiments are presented in Fig. 10.8. The experimental design was identical to that for the experiments with EMS. The amount of MH ranged from 5.0×10^{-3} to 0.5 μmole per plant. The frequency of mutant pollen grains ranged from 4.59×10^{-5} for the control to 5.54×10^{-4} for the highest-treatment group. A statistically significant increase in the frequency of mutant pollen grains was observed for all treatment groups. This increase was linearly concentration-dependent ($r^2 = 0.99$). The percentage of pollen abortion increased directly with the concentration of MH ($r = 0.96$). Thus, MH induces gametophytic death at concentrations that are highly mutagenic. These data clearly indicate that MH is an exceedingly potent mutagen in maize.

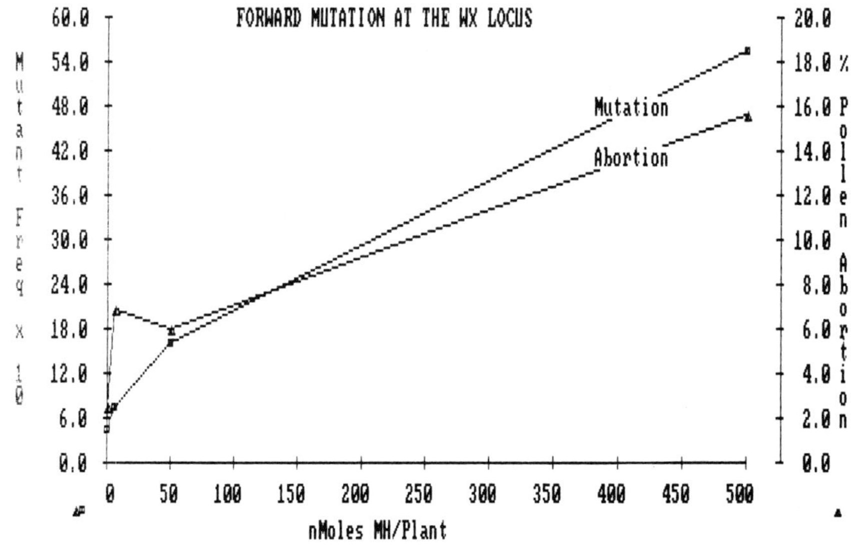

Fig. 10.8 – Concentration-response curve for chronic exposure of Early–Early Synthetic maize to maleic hydrazide. Mutation represents the induction of forward *wx* mutant pollen grains in plants that were *Wx/Wx*. Abortion indicates the percentage of pollen abortions.

10.5.2 Analysis of municipal sewage sludge with Early–Early Synthetic

Forward mutation at the *wx* locus was assayed after exposure to extremely high concentrations of Chicago municipal sewage sludge [10]. The sludge was from the lot used in the *in situ* test plots already described. The experiment involved three treatment groups of five plants each. Each treatment group differed only in the ratio of sludge to soil mixture and in the dilution of the sludge used to water the plants. The highest-treatment group was planted in a mixture of one part sludge with two parts standard soil-mix. These plants were watered with a threefold dilution of sludge with demineralized water. The middle-treatment group was planted in a 1:5 sludge–soil mixture and watered with a sixfold dilution of sludge. The low-treatment group was planted in 1:11 sludge-soil mix and watered with a twelvefold sludge dilution. The concurrent control of sibling kernels used standard soil and pure water. The plants were grown in a plant-growth chamber and the tassels were harvested and coded, and the pollen grains were analysed for the frequency of forward mutant pollen grains and pollen abortion.

The results are presented in Fig. 10.9. They indicate that the administration of sludge increased the frequency of *wx* mutant pollen grains by two orders of magnitude. The control frequency of mutant pollen grains of 4.31×10^{-5} was consistent with the spontaneous frequency for this variety (see Figs. 10.7 and 10.8). The frequencies of mutant pollen grains decreased with increasing amounts of added sludge, indicating that even the lowest concentration of applied sludge was high enough to reach the toxic region of the mutational concentration-response curve. The percentage of pollen abortions was higher at the higher sludge concentrations (19.9%) than for the control group (13.5%). The data indicate that the sludge induced a significant increase in forward mutation in the pollen grains. The increased pollen abortion rate indicated the presence of toxic as well as mutagenic agents.

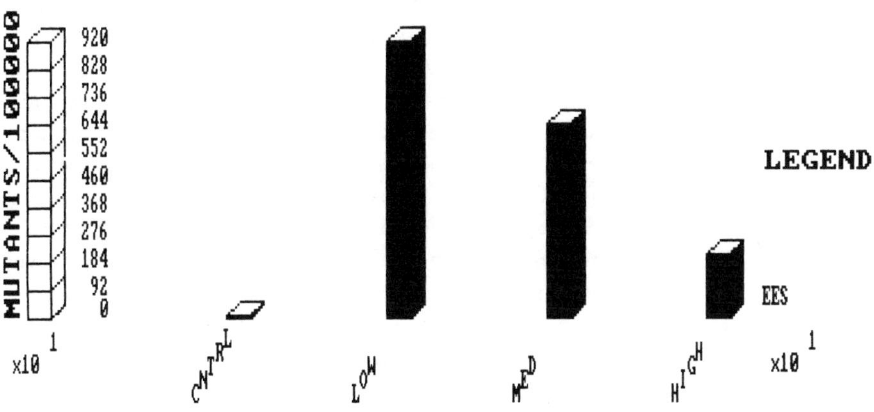

Fig. 10.9 – The induction of forward mutant *wx* pollen grains by Chicago municipal sewage sludge in Early–Early Synthetic maize plants that were *Wx/Wx*. The bars represent the frequency of mutant pollen grains per treatment group. The abbreviations of the treatment groups are: CNTRL = Control, LOW = 1/12 sludge concentration, MED = 1/6 sludge concentration, HIGH = 1/3 sludge concentration.

10.5.3 Additional genetic end points in maize

Two sensitive genetic end points are now being introduced and calibrated for use with Early-Early Synthetic. They are forward mutation at the *yg2* locus and the induction of micronuclei in root-tip cells. Although these assays have not been used for *in situ* analysis of environmental mutagens, laboratory experiments demonstrate their potential for use in the field.

Forward Mutation at the yg2 *Locus in Maize.* The *yg2* locus is on the short arm of chromosome 9. Homozygous *yg2* plants are pale yellow-green, whereas heterozygous plants (*Yg2/yg2*) have normal green leaves. However, if the dominant allele is lost owing to a gene mutation (a chromosome aberration that leads to the deletion of the locus or to the loss of the entire chromosome that carries *Yg2*), the recessive allele is expressed and the cell is deficient in the production of chlorophyll. When a mutant cell divides, a sector or clone of mutant progeny cells is produced. If a leaf primordial cell suffers a loss of the dominant allele the resulting leaf will be a chimera and the clone of *yg2* mutant cells will appear as a yellow-green sector within the green leaf. The advantages of this assay are many. It detects forward mutation at a specific locus in somatic cells of a higher eukaryote. It is a quantitative assay with high sensitivity. The large number of target cells provides a high degree of genetic resolution. The assay is inexpensive and relatively rapid, for a test based on an intact higher eukaryotic organism.

Heterozygous *yg2* kernels that contained 50% Early–Early Synthetic germ-plasm were prepared by appropriate genetic crosses. These kernels were used to analyse the kinetics of mutation induction by EMS and gamma-rays. The kernels were soaked in aerated distilled water for 72 hr at 20°C. EMS solutions in buffer were freshly prepared for each experiment, at concentrations ranging from 0.25 to 20.00m*M*. The prepared kernels were exposed to the EMS solution for 8 hr at 20°C, under aeration. A concurrent control, in which kernels were exposed to a 0.1*M* phosphate buffer (pH 7.4), was conducted with each experiment. After treatment the kernels were washed and planted in standard soil and grown in a plant-growth chamber at 20°C until the muturation of leaves four and five.

The control and treatment groups were coded, and when leaves four or five had matured (identified by the presence of the ligule) they were scored. Ultra-violet light and a magnifying lens were used as aids in scoring the leaves. Only sectors that were yellow-green and greater than 1 mm in length were counted (Fig. 10.10). The number of sectors per leaf was recorded and when the entire experiment had been scored, the groups were decoded [21].

The results of the EMS experiments are presented in Fig. 10.11. The data from individual experiments were combined and the aggregated data were analysed. The mean number of *yg2* sectors per leaf four ranged from 0.19 for the control to 12.85 for the 20m*M* treatment group. The induction of forward mutants in leaf-four cells was linear ($r^2 = 0.96$) and concentration-dependent. The mean number of *yg2* sectors per leaf five ranged from 0.11 for the control to 2.78 for the 20m*M* treatment group. As with leaf four, the induction of forward mutation in leaf five was linear ($r^2 = 0.86$) and concentration-dependent.

Kernels used in the gamma-radiation experiments were prepared as outlined above. The kernels were placed in polyethylene tubes and exposed to gamma-rays in a Shepard irradiator that contained a ^{137}Cs 2,500-Ci source. The doses of gamma radiation ranged from 25 to 500 rads. A concurrent control that was not irradiated was conducted with each experiment. The kernels were grown and cultivated as outlined above.

The results of the gamma-radiation treatments are presented in Fig. 10.12. The data from all the experiments were combined and the aggregated data

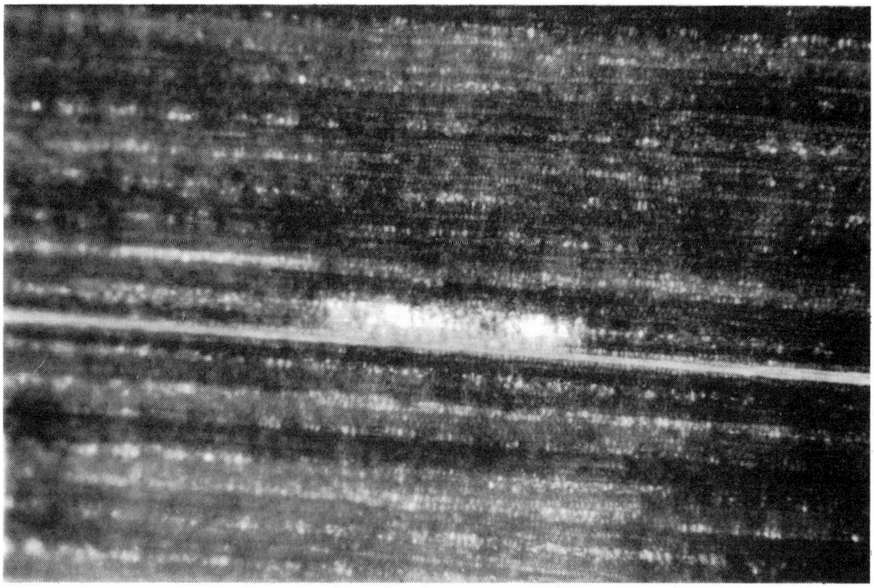

Fig. 10.10 – A *yg2* sector in leaf four.

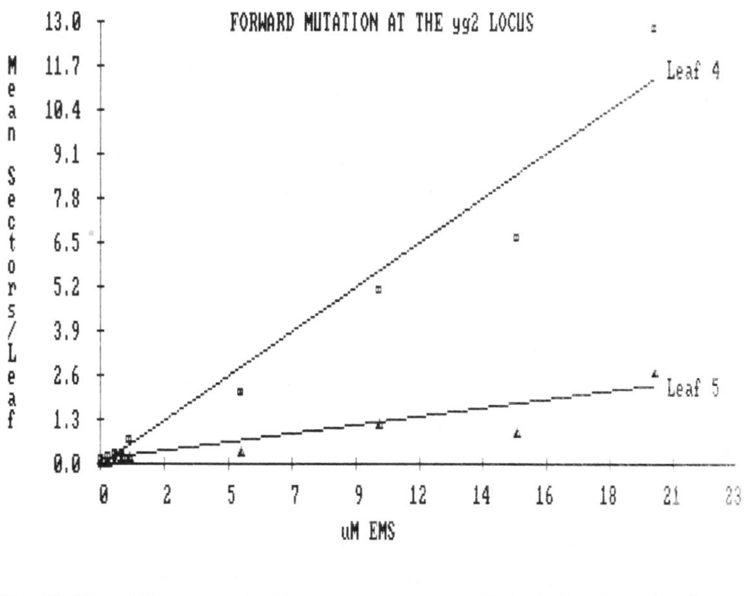

Fig. 10.11 – The concentration-response curve of the induction of *yg2* sectors in leaves four and five by ethyl methanesulphonate.

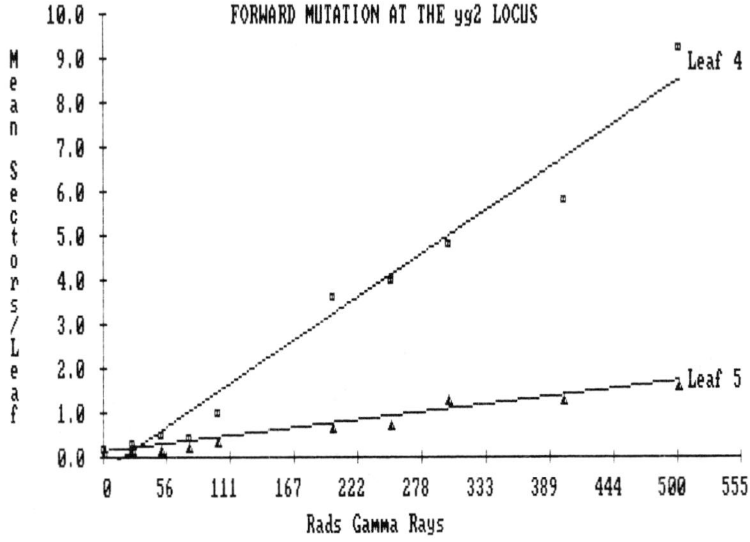

Fig. 10.12 – The dose-response curve of the induction of *yg2* sectors in leaves four and five by gamma radiation.

analysed. The mean numbers of sectors per leaf four for the control and the 500-rad treatment group were 0.18 and 9.26, respectively. The concentration-response curve was linear ($r^2 = 0.97$). The data for leaf five showed the mean number of sectors per leaf for the control was 0.11, and for the highest radiation treatment group 1.61. The induction of forward mutation in leaf five was linear ($r^2 = 0.96$) and concentration-dependent.

These data demonstrate that the induction of forward mutation at the *yg2* locus in maize leaf four or five primordial cells is a quantitative and sensitive genetic assay. Although these lines have not been used to monitor for environmental pollution it is evident that the assay may be used in such a manner.

Induction of Micronuclei in Maize Root Tip Cells. An assay to detect chromosome aberrations in individual plants that are used in the *wx* or *yg2* point mutation assays has been devised [22]. The assay is based on the induction of micronuclei in root-tip cells. Micronuclei are formed from acentric fragments, lagging chromosomes or multicentric chromosomes connected by bridges (Fig. 10.13). An experiment was conducted to determine whether ethyl methanesulphonate could induce micronuclei in maize root-tip cells. Kernels heterozygous for the *yg2* locus were prepared as described above. The kernels were treated for 8 hr at 20°C, under aeration, in solutions of EMS. The concentrations tested were zero (control), 1m*M*, 10m*M* and 20m*M* EMS. Following treatment, the kernels were rinsed and planted in standard soil. The pots were placed in a plant-growth chamber and grown for 5 days. Root tips were cut from the plants and fixed for 24 hr in ethanol–acetic acid mixture (3:1 v/v). The root tips were rinsed in demineralized water, then placed in 1*M* hydrochloric acid at 60°C for 7.5 min.

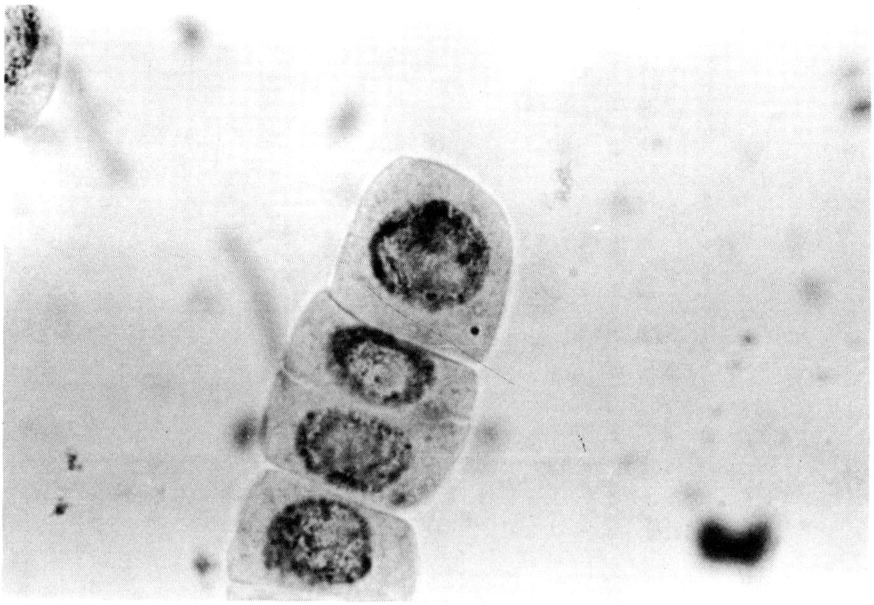

Fig. 10.13 – A micronucleus in a maize root-tip cell.

The root tips were rinsed again and placed in Feulgen stain for 1 hr, followed by treatment with 5% pectinase solution for 1 hr. Each slide was prepared by squashing one root tip in the stain. The number of interphase cells counted on each slide was approximately 1000 and five slides were analysed for each control or treatment group. The results are presented in Fig. 10.14. The micronucleus frequencies for all treatment groups were significantly higher than that for the control. The concentration of EMS and the induction of micronuclei are directly related.

The advantages of the micronucleus test in maize root-tip cells for analysis of environmental pollutants are many. It is a relatively easy and rapid assay that can be conducted under acute or chronic exposure regimens. It is sensitive to acute treatment with a classical mutagen, EMS, and with appropriate genetic stocks it can be used simultaneously with the *wx* and *yg2* point mutation assays in maize.

10.6 CONCLUSIONS

The use of plant species as one of a battery of genetic indicator organisms for the study and evaluation of contaminated areas is recommended. The data presented in this paper are sufficient to demonstrate the sensitivity of specially designed maize lines for environmental monitoring for pollution control.

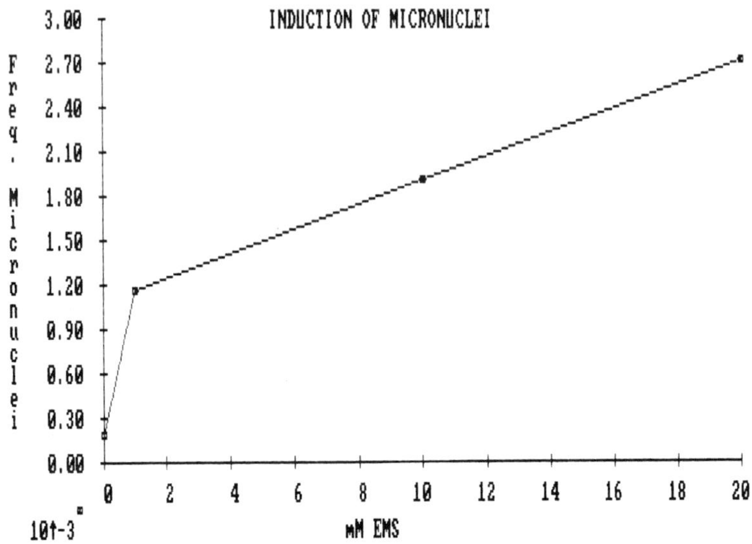

Fig. 10.14 – The concentration-response curve of the induction of micronuclei in maize root-tip cells by ethyl methanesulphonate. Each division on the ordinate is equal to 0.30×10^{-3}.

REFERENCES

[1] W. R. Lower, P. S. Rose and V. K. Drobney, *Mutation Res.*, 1978, **54**, 83.
[2] T. H. Ma, *Environ. Health Perspectives*, 1981, **37**, 85.
[3] L. A. Schairer and R. C. Sautkulis, in *Environmental Mutagenesis, Carcinogenesis, and Plant Biology*, Vol. 2, E. J. Klekowski, Jr. (ed.), pp. 155–194, Praeger Publishers, New York, 1982.
[4] E. J. Klekowski, Jr., *Environmental Mutagenesis, Carcinogenesis, and Plant Biology*, Vols. 1 and 2, Praeger Publishers, New York, 1982.
[5] D. D. DeMarini, M. J. Plewa and H. E. Brockman, *J. Tox. Environ. Health*, 1982, **9**, 127.
[6] M. W. Heartlein, D. M. DeMarini, A. J. Katz, J. C. Means, M. J. Plewa and H. E. Brockman, *Environ. Mutagenesis*, 1981, **3**, 519.
[7] M. J. Plewa and E. D. Wagner, *Environ. Health Perspectives*, 1981, **37**, 61.
[8] J. M. Gentile, G. J. Gentile, J. Bultman, R. Sechriest, E. D. Wagner and M. J. Plewa, *Mutation Res.*, 1982, **101**, 19.
[9] M. J. Plewa, E. D. Wagner, G. J. Gentile and J. M. Gentile, *Mutation Res.*, 1984, **136**, 233.
[10] P. K. Hopke, M. J. Plewa, J. B. Johnston, D. Weaver, S. G. Wood, R. A. Larson and T. Hinesly, *Environ. Sci. Technology*, 1982, **16**, 140.
[11] M. J. Plewa and P. K. Hopke, in *Carcinogens and Mutagens in the Environment*, Vol. 3, H. F. Stich (ed.), pp. 155–175, CRC Press, Boca Raton FL, 1983.
[12] S. Ichikawa, *Environ. Health Perspectives*, 1981, **37**, 145.
[13] R. A. Nilan, *Environ. Health Perspectives*, 1978, **27**, 181.
[14] J. F. Crow and S. Abrahamson (Chairmen), *Identifying and Estimating the Genetic Impact of Chemical Mutagens*, National Academy Press, Washington, D. C., 1982.
[15] J. F. Crow, *Environ. Health Perspectives*, 1973, No. 6.
[16] E. P. Benditt, *Scientific Am.*, 1977, **236**, 74.
[17] B. N. Ames, *Science*, 1979, **204**, 587.
[18] M. Sorsa, *J. Tox. Environ. Health*, 1980, **6**, 57.

[19] M. J. Plewa and J. M. Gentile, *Final Report* EPA-600/S1-83-006, U.S. Environmental Protection Agency Health Effects Research Laboratory, Research Triangle Park, NC, 1983.
[20] W. R. Engels, *Environ. Mutagenesis,* 1979, **1**, 37.
[21] P. A. Dowd and M. J. Plewa, *Maize Genet. Coop. News Lett.,* 1983, **57**, 144.
[22] E. D. Wagner and M. J. Plewa, *Maize Genet. Coop. News Lett.,* 1983, **57**, 141.

11

Induction of mitotic aneuploidy in yeast with aprotic polar solvents

F. K. ZIMMERMANN, V. W. MAYER*, R. E. TAYLOR-MAYER†, U. GRÖSCHEL-STEWART‡ and I. SCHEEL, Institut für Mikrobiologie, Technische Hochschule, D-6100 Darmstadt, Federal Republic of Germany

11.1 INTRODUCTION

The vast majority of the commonly used mutagenicity tests detect point mutattions, mitotic recombination, chromosomal structural aberrations such as chromosome breaks or translocations, and sister-chromatid exchanges, or are based on certain other reactions initiated by chemical interactions with DNA or DNA metabolizing enzymes [1].

Proper segregation of chromosomes in mitotic and meiotic nuclear divisions not only requires intact chromosomes but also an intact spindle fibre apparatus, the structure which pulls apart the replicated daughter chromosomes in mitosis, or separates the synapsed homologous chromosome pairs in the first meiotic division. Malfunction of this structure can result in irregular, aneuploid chromosome numbers. The important difference between the induction of chromosomal mal-segregation leading to aneuploid chromosome numbers and induction of other types of mutagenicity (e.g. point mutation, mitotic recombination, chromosomal structural aberrations and sister-chromatid exchanges) is that the target molecules for induction are not the same. Agents specifically interfering with the function of the spindle fibre apparatus may induce aneuploidy but may not induce other types of genetic changes such as point mutation or chromosome breaks. Consequently some agents which induce aneuploidy by interfering with the components of the spindle fibre apparatus have been classified as non-mutagens because they have been tested only for other end points.

Permanent addresses: *Genetic Toxicology Branch, Division of Toxicology, Food and Drug Administration, Washington, DC; †Department of Biology, Slippery Rock University, Slippery Rock, PA 16057, U.S.A.; ‡Institut für Zoologie, Technische Hochschule, Darmstadt.

Parry *et al.* [2] reported on the induction of mitotic aneuploidy by tumour promoters which did not induce any other types of genetic effects. We have found that aprotic polar solvents such as acetone, organic esters, and certain ketones strongly induce mitotic aneuploidy in yeast, but do not induce point mutation or mitotic recombination [3]. This demonstrates the need for tests which detect induction of aneuploidy.

Induction of mitotic aneuploidy by aprotic polar solvents in the fungus yeast does not necessarily mean that the same agents will exert the same effect in man. Himes *et al.* [4] demonstrated, however, that the aprotic polar solvent dimethylsulphoxide accelerated the aggregation of mammalian tubulin to form microtubules. We also demonstrated that aprotic polar solvents had an effect on the assembly *in vitro* of porcine brain tubulin. Dimethylsulphoxide and acetone accelerated assembly; methyl ethyl ketone, ethyl acetate and acetonitrile inhibited assembly [5]. These findings suggest that aprotic polar solvents may induce aneuploidy in mammals.

Aneuploidy in man contributes to spontaneous abortions [6, 7] and is the cause of well defined genetic syndromes such as Down's (trisomic condition for chromosome 21) Turner's (only one X chromosome) or Klinefelter's (two X chromosomes and one Y chromosome).

11.2 MATERIALS AND METHODS

11.2.1 The test system

The diploid yeast strain D61.M was bred by Zimmermann and Scheel [8]. The genotype is:

$$\frac{\mathbf{cyh2}\quad +\quad \mathbf{leu1}}{+\quad \mathbf{trp5}\quad +}\ \text{centromere}\ \frac{\mathbf{ade6}}{+}\ \frac{\mathbf{ade2}}{\mathbf{ade2}}\ \frac{\mathbf{ilv1\text{-}92}}{\mathbf{ilv1\text{-}92}}$$

$$\text{centromere}\ \frac{\mathbf{MATa}\qquad \mathbf{MAL2}}{\mathbf{MATalpha}\quad \mathbf{mal}}\ \frac{\mathbf{SUC}}{\mathbf{suc}}\ \frac{\mathbf{HIS}}{\mathbf{his}}$$

The adenine requirement resulting from the homozygous condition of **ade2** causes strain D61.M to form red colonies on media with a limited amount of adenine. The homozygous condition of **ilv1-92** causes a requirement for isoleucine. This strain is inhibited by 1.5 ppm cycloheximide because the resistance-conferring allele **cyh2** is recessive. Alleles **ade6, his, leu1** and **trp5** are recessive mutant alleles which when expressed cause requirements for adenine, histidine, leucine and tryptopham respectively. The appearance of white colonies is indicative of the expression of **ade6.** Expression of the other three markers can only be recognized by a lack of growth on synthetic media containing no histidine, leucine or tryptophan, whereas expression of **mal** and **suc** is detected by a lack of growth on media with 2 ppm antimycin A, and maltose or sucrose as a fermentable carbon source instead of the usual glucose supplement. The normal diploid cells of strain D61.M cannot mate when brought into contact with haploid cells carrying one of the two alleles of the mating type locus. Elimination of one of these alleles from a diploid cell will result in the expression of a mating type function as determined by the remaining allele.

11.2.2 Media

The rich medium consisted of 2% peptone, 2% glucose and 1% yeast extract (YEPD). Yeast extract and peptone were also the basis for the diagnostic media used to test for the ability to utilize maltose and sucrose in the presence of 2 ppm antimycin A. A YEPD medium with 1.5 ppm cyclohexamide was used to detect expression of recessive **cyh2**. The synthetic medium used consisted of 1.7 g/l. DIFCO yeast nitrogen base without amino-acids and ammonium sulphate, and was supplemented with 5 g/l. ammonium sulphate and 2% glucose. For a full development of the red colony pigment in **ade2** homozygotes only 4 mg/l. adenine was used instead of the usual supplement of 10 mg/l. The amino-acid supplement included 10 mg/l. L-histidine, 60 mg/l. L-isoleucine, 30 mg/l. L-leucine and 10 mg/l. L-tryptophan. Individual supplements were omitted from media diagnostic for the expression of the different recessive nutritional markers. Media were solidified with 10 g/l. Oxoid agar.

11.2.3 Preparation of cultures

The frequency of resistant colonies expressing only the recessive resistance marker **cyh2** usually varied between 20 and 40 per 10^5 cells. Much higher frequencies, however, are observed in about one out of five cultures. Aneuploids expressing **cyh2, ade6** and **leu1** are quite rare, usually occurring at frequencies between 0.2 and 2.0×10^{-6}. Therefore it was necessary to plate several million cells in order to determine the spontaneous background of white resistant and leucine-requiring aneuploids. A culture with an unusually high titre of resistant colonies would lead to an overcrowding of the selective cycloheximide plates. A special procedure was followed in order to prepare cultures with a normal background of resistant cells. Approximately 200 cells of strain D61.M were inoculated into 5-ml portions of liquid YEPD medium in metal-capped test-tubes and incubated at 28°C for about 36–40 hr on a reciprocal shaker. At this time the cell titre was approximately 5×10^6 cells/ml. Samples of 0.1 ml were plated onto the selective cycloheximide medium and incubated for 5-6 days until the colonies had grown and developed enough pigment to distinguish between the normal red and the very rare white colonies expressing **ade6**. The culture tubes were kept in a refrigerator and used for up to three weeks. Cultures were diluted 1:10 or 1:20 with fresh liquid YEPD medium before each experiment and incubated for 2–4 hr at 28°C on a reciprocal shaker until the chemicals were added. This was sufficient to bring the cells back into the growth phase.

11.2.4 Chemicals

Dimethylsulphoxide, ethyl acetate and pyridine (from Merck, Darmstadt) were analytical grade. Acetonitrile was a gift from Dr. R. Schlegelmilch, Institute of Pharmacology, University of Ulm, West Germany; benzonitrile, octanonitrile, phenylacetonitrile and *o*-toluonitrile were gifts from Dr. Gais, Department of Organic Chemistry of our university. All were at least 98% pure.

11.2.5 Treatment procedure

In order to detect chemicals which induce chromosomal mal-segregation, all treatments were performed with growing cells in a growth medium. The solvents were added directly to a constant volume of 2 ml of a liquid YEPD culture in metal-capped 10-ml centrifuge tubes 16 mm in diameter. Cell titres at the time of addition of the chemicals were between 3 and 7 x 10^6 cells per ml. Incubation for 4 hr was performed at 28°C on a reciprocal shaker, followed by 16-17 hr overnight storage in an ice-bath. The cultures were returned to the shaker at 28°C for another 4 hr. Cell titres were then determined by haemocytometer counting and the cultures were diluted if necessary to a titre of 20-30 x 10^6 cells per ml and directly plated on the selective cycloheximide medium. Further dilutions were also plated on a non-selective synthetic medium to determine the titre of colony-forming units. All plates were incubated at 28°C. Colonies on the non-selective medium were counted after four days; the selective cycloheximide plates were scored after six days. All white colonies on the selection plates were then tested for a leucine requirement. This additional test was omitted only in the case of the positive control agents ethyl acetate and acetonitrile since the frequency of white resistant colonies was usually very high and it had been established in previous experiments that such colonies were due to the induction of mitotic aneuploidy [3].

All experiments tested each chemical in a series of increasing concentrations with very small dose increments, and compared the frequencies of aneuploidy with those of both a negative control and a positive control with either ethyl acetate or acetonitrile, agents known to induce mitotic aneuploidy in strain D61.M.

11.2.6 Evaluation of data

The colony counts on three plates of non-selective medium were used to compute the titres of colony-forming units in the cultures at the time of plating. These in turn were used to compute the relative frequencies of all resistant colonies (both red and white) and those of the leucine-requiring white resistant colonies. The latter category was considered to reflect induction of chromosomal mal-segregation in those cases in which the relative frequency of red resistant colonies did not increase. A lack of increase in the red resistant colonies indicates that the treatment did not induce mitotic recombination, point mutation or deletion of chromosomal fragments.

11.2.7 Presentation of data

The results of the experiments are presented in Tables 11.1-11.4. The first column gives the concentrations of the chemical being tested: the titre of colony-forming units is given in the second column; the actual counts of all resistant colonies on the selective medium (from three plates unless otherwise stated) are given in the third column, and the relative frequencies of all resistant colonies per 10^5 colony-forming units are given in the fourth column. The figures in the fourth column demonstrate the cumulative effects of all genetic events which lead to the expression of the recessive resistance marker **cyh2**. The actual numbers of the leucine-requiring white resistant colonies (also usually from three plates) and the relative frequencies per 10^6 colony-forming units are found in the last two columns. These figures represent those events of chromosomal mal-segregation which lead to chromosome loss.

Table 11.1 – Induction of Mitotic Aneuploidy in Yeast Strain D61.M with Dimethylsulphoxide

Sample and concentration	Titre $\times 10^6$	Total Resistants Colonies	Total Resistants $\times 10^{-5}$	Presumptive Monosomics Colonies	Presumptive Monosomics $\times 10^{-6}$
Control	37.3	980†	52.6	3	0.54
Positive control:acetonitrile					
3.85%	16.9	929†	54.9	380*	74.8
Dimethylsulphoxide					
6.54%	28.3	2058	48.4	7	1.65
7.41%	20.0	976†	48.9	16	2.67
8.26%	13.7	789†	57.7	10	2.44
9.09%	11.9	1619	45.5	16	4.48
9.91%	7.3	1388	63.4	40	18.3

Titres were determined at the time of plating and are based on the determination of colony-forming units. Total resistants refers to the numbers of colonies – both red and white – on the selective cycloheximide medium. Presumptive monosomics were those white resistant colonies on the selective medium which were tested and found to express a leucine requirement. White colonies not expressing this requirement are not included. †Only one plate counted instead of three. *These colonies were not tested for a leucine requirement since acetonitrile is an established inducer of mitotic aneuploidy in yeast.

Table 11.2 – Induction of Mitotic Aneuploidy in Yeast Strain D61.M with Pyridine

Sample and concentration	Titre $\times 10^6$	Total Resistants Colonies	Total Resistants $\times 10^{-5}$	Presumptive Monosomics Colonies	Presumptive Monosomics $\times 10^{-6}$
Control	156.3	1046	22.3	1	0.21
Positive control: ethyl acetate					
2.44%	2.7	385	47.0	191*	232.9
Pyridine					
0.70%	16.5	507	20.4	3	1.21
0.79%	11.9	860	24.0	15	4.19
0.89%	9.1	757	27.7	26	9.55
0.99%	6.1	625	33.4	28	15.0
1.08%		535	33.4	27	
1.19%	5.1	566	37.0	31	20.2
1.28%	4.4	385	28.1	36	27.1

*These white colonies were not tested for a leucine requirement since ethyl acetate is an established inducer of mitotic aneuploidy in yeast.

Table 11.3 – Induction of Mitotic Aneuploidy in Yeast Strain D61.M with γ-Butyrolactone and γ-Valerolactone

Sample and concentration	Titre $\times 10^6$	Total Resistants		Presumptive Monosomics	
		Colonies	$\times 10^{-5}$	Colonies	$\times 10^{-6}$
Control	88.5	1847	34.8	5	0.95
γ-Butyrolactone					
2.44%	28.2	1668	39.4	13	3.07
2.91%	9.1	959	35.0	20	7.30
3.38%	5.3	572	35.8	75	47.2
Control	51.5	1879	36.7	0	–
γ-Valerolactone					
1.48%	57.7	2930	33.8	2	0.23
1.72%	26.6	2801	35.1	7	0.88
1.96%	19.0	1968	34.5	10	1.75
2.20%	16.5	1788	36.1	17*	5.15
2.44%	13.4	1485	36.9	21	5.21
2.68%	8.4	1050	41.8	49	19.5
2.91%	5.7	601	34.9	17	9.88

*Only two plates scored instead of three.

Table 11.4 – Induction of Mitotic Aneuploidy in Yeast Strain D61.M with Different Nitriles

Sample and concentration	Titre $\times 10^6$	Total Resistants		Presumptive Monosomics	
		Colonies	$\times 10^{-5}$	Colonies	$\times 10^{-6}$
Control	64.7	1645	33.9	17	3.51
Acetonitrile					
4.31%	9.6	690	24.0	259	89.9
Control	11.0	1014	30.7	1	0.30
Positive control: ethyl acetate					
2.44%	8.3	673	27.1	121*	48.8
n-Butyronitrile					
0.25%	7.0	1050	50.2	2	0.96
0.50%	6.2	1152	61.9	1	0.54
0.62%	4.9	869	59.5	1	0.68
0.74%	5.0	1454	96.2	2	1.32
0.85%	5.4	1053	64.6	1	0.61
0.99%	4.1	922	75.0	0	–
1.11%	3.4	979	96.9	1	0.99
1.23%	2.8	725	86.3	1	1.19
1.48%	1.6	278	57.9	0	–
Control	62.9	1767	28.1	1	0.16

Table 11.4 – *continued*

Sample and concentration	Titre × 10^6	Total Resistants Colonies	× 10^{-5}	Presumptive Monosomics Colonies	× 10^{-6}
Positive control: ethyl acetate					
1.96%	7.5	1151	51.4	494*	220.5
n-Octanonitrile					
0.099%	47.8	1932	27.0	2	0.28
0.148%	29.8	2318	25.9	5	0.56
0.196%	18.4	1422	25.8	1	0.18
0.244%	6.0	567	31.7	1	0.56
Control	55.9	1478	22.0	2	0.30
Positive control: ethyl acetate					
2.20%	1.0	287	99.0	202*	696.5
Benzonitrile					
1.36 μl/ml	23.4	1617	23.0	6	0.85
1.48 μl/ml	17.8	1544	28.9	6	1.12
1.60 μl/ml	13.4	1187	29.5	6	1.49
1.72 μl/ml	10.1	1037	34.1	8	2.63
1.84 μl/ml	7.7	811	35.2	3	1.30
1.96 μl/ml	5.7	716	41.9	8	4.68
Phenylacetonitrile					
1.19 μl/ml	24.7	2074	28.0	8	1.08
1.39 μl/ml	11.1	1263	38.0	13	3.91
1.59 μl/ml	6.4	794	41.1	42	21.8
Control	51.2	1879	36.7	0	–
Positive control: γ-valerolactone					
2.68%	8.4	1050	41.8	49	19.5
o-Toluonitrile					
0.68 μl/ml	17.0	1753	34.4	2	0.39
0.74 μl/ml	7.6	962	42.0	8	3.49
0.80 μl/ml	6.8	761	36.8	5	2.45
0.86 μl/ml	5.9	697	39.6	10	5.68

*White colonies not tested for a leucine requirement.

A detailed genetic analysis of red and white resistant colonies resulting from cultures exposed to several agents has been presented by Zimmermann *et al.* [3].

11.2.8 Preparation of porcine brain tubulin

Pig brains were obtained immediately after slaughtering from a local butcher and processed at once as previously described by Gröschel-Stewart *et al.* [5]. The purification method (Williams and Lee [9]) is based on the ability of microtubules to dissociate in the cold into the individual monomers of tubulin and microtubule-associated proteins and to reassociate at temperatures above 17°C. Twice recycled preparations contained 16 mg of protein per ml [10] and were 90% pure.

11.2.9 Assembly of microtubules *in vitro*

All ingredients were kept at ice-bath temperature and mixed in a cuvette to give a total volume of 1.00 ml. Assembly buffer, solvent chemical, 250–300 μl of the tubulin preparation, and 10 μl of a 50m*M* solution of GTP were mixed in the cuvette, and tubulin assembly was followed by means of a Beckman spectrophotometer, at 32°C. The increase in absorbance at 350 nm was recorded for three experiments in parallel, measured against a control cuvette without solvents. The figures shown are direct recordings from the photometer.

11.3 RESULTS

Ethyl acetate and acetonitrile were used as positive control agents because they are strong inducers of mitotic aneuploidy (as can be seen from Tables 11.1, 11.2 and 11.4). Both agents affect tubulin aggregation *in vitro* (Figs. 11.1 and 11.2).

Himes *et al.* [4] showed that 10% dimethylsulphoxide accelerated the aggregation of bovine brain tubulin *in vitro.* We had previously observed that two other aprotic polar solvents, acetone and 2-methoxyethyl acetate, accelerated aggregation of porcine brain tubulin *in vitro* and strongly induced mitotic aneuploidy in yeast strain D61.M [3, 5]. This observation prompted us to test dimethylsulphoxide (DMSO) for aneuploidy induction. Chromosomal mal-segregation was induced at a DMSO concentration of approximately 10% (Table 11.1). Although there was a significant induction of aneuploidy, there was no parallel increase in red resistant colonies, indicating that aneuploidy was the only genetic effect induced by dimethylsulphoxide.

Pyridine is a polar aprotic solvent frequently used in chromatography. It induced chromosomal mal-segregation in strain D61.M in a dose-related fashion, but again there was no indication of the induction of other types of genetic effects (Table 11.2).

Lactones also belong to the group of aprotic polar solvents. We tested γ-butyro- and γ-valerolactone. As shown in Table 11.3, there was a strong induction of mitotic aneuploidy with γ-butyrolactone (Table 11.3) and a somewhat weaker effect with γ-valerolactone (Tables 11.3 and 11.4). There was no evidence with either lactone for the induction of other types of genetic effects.

n-Butyronitrile did not induce mitotic aneuploidy, but a threefold increase in red resistant colonies indicated induction of other types of genetic events such as point mutation, mitotic recombination and/or deletion of chromosomal fragments (Table 11.4, second experiment).

n-Octanonitrile was toxic to yeast cells, but it did not induce any genetic effects at the concentrations tested (Table 11.4, third experiment).

Benzonitrile may be a weak inducer of mitotic aneuploidy, but the numbers of white leucine-requiring resistant colonies are too small to allow any firm conclusions. A small increase in the red resistant colonies may reflect a possible induction of point mutation, mitotic recombination and/or deletion of chromosomal fragments (Tables 11.4, fourth experiment).

o-Toluonitrile may also be a weak inducer of mitotic aneuploidy but the effects were slight and the numbers of white leucine-resistant colonies were small (Table 11.4, sixth experiment). There was no indication of other types of genetic events.

Phenylacetonitrile, however, was a strong inducer of mitotic aneuploidy and a weak inducer of point mutation, mitotic recombination and/or deletion of chromosome fragments (Table 11.4, fifth experiment). The same pattern was observed in three experiments not reported here. Phenylacetonitrile was the most effective inducer of mitotic aneuploidy among all the polar aprotic solvents tested in yeast in this and in previous investigations [3, 11].

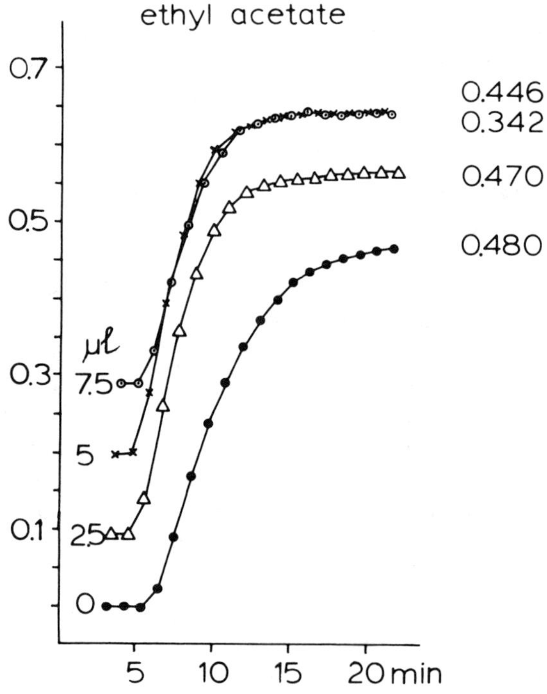

Fig. 11.1 – The effects of ethyl acetate on the assembly *in vitro* of twice recycled porcine brain tubulin.

A 300-μl volume of a preparation of twice recycled porcine brain tubulin in a total volume of 1.00 ml was caused to aggregate in assembly buffer at 32°C in the absence and presence of various amounts of analytical grade ethyl acetate. Aggregation was followed by recording of the increase in absorbance at 350 nm as a function of time. The points on the curves are directly taken from the absorbance recordings. The volumes of ethyl acetate contained within the 1.00 ml volume are indicated at the lower end of each curve; the values at the upper ends of the curves represent the total absorbance increments. The ordinate gives the relative increase in absorbance; the different initial absorbance levels in the various samples were caused by setting different base levels for each of the four cuvettes tested in parallel.

The effects of ethyl acetate and acetonitrile on the assembly *in vitro* of twice recycled porcine brain tubulin are shown in Figs. 11.1 and 11.2. Ethyl acetate at 0.75% reduced the extent of tubulin assembly to 71% and acetonitrile at 0.1% reduced it to 82%, relative to the concurrent control assays. Still higher concentrations caused a complete inhibition, as reported previously [5].

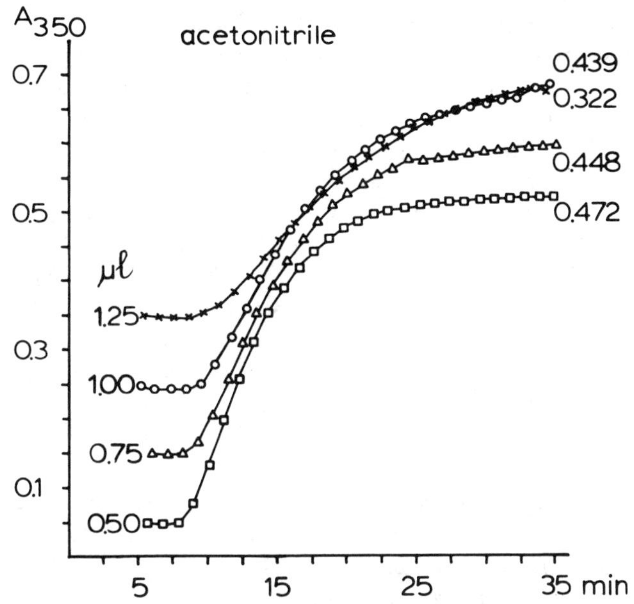

Fig. 11.2 – The effects of acetonitrile on the assembly *in vitro* of twice recycled procine brain tubulin.
For procedure see legend on Fig. 11.1 (read acetonitrile instead of ethyl acetate!).

11.4 DISCUSSION

The aprotic polar solvents dimethylsulphoxide, pyridine, γ-butyrolactone, γ-valerolactone and acetonitrile specifically induced mitotic aneuploidy with no indication of the induction of point mutation, mitotic recombination and/or deletion of chromosomal fragments. Phenylacetonitrile was a strong inducer of mitotic aneuploidy and a weak inducer of other types of genetic events; benzonitrile weakly induced aneuploidy and other types of genetic events. *o*-Toluonitrile was at best a weak inducer of aneuploidy alone. n-Butyronitrile did not induce aneuploidy but showed effects in the categories of mitotic recombination, point mutation and deletion of chromosomal fragments. n-Octanonitrile was totally non-mutagenic at the concentrations tested.

These results together with results published elsewhere [3, 11] clearly demonstrate that aprotic polar solvents can be strong inducers of aneuploidy without inducing any other type of mutation [3]. The most potent solvent with such a specific mutagenic activity is ethyl acetate, used as a positive control agent in the present investigation. Most of the aprotic polar solvents used are not very reactive chemically under physiological conditions and their ability to induce mitotic aneuploidy varies strongly. The mechanism of action in aneuploidy induction must thus be different from that for induction of other types of genetic change.

Since white cycloheximide-resistant and leucine-requiring colonies can be found in experiments where there is no indication of induction of any other types of genetic events, the presence of this category of aberrant colony types could not result from multiple events of point mutation, mitotic recombination or deletion of chromosomal fragments. Such events would lead to an increase in red resistant colonies, and this was not found. Genetic analysis of white, cycloheximide-resistant and leucine-requiring colonies in platings from control and treated cultures revealed that most of them were completely diploid; only 7 out of 52 tested showed the lethal segregation which was to be expected after loss of one chromosome from a diploid yeast strain. The analysis revealed that 5 out of 52 had become trisomic for chromosome III [3].

The test was designed to detect monosomic cell clones. Non-disjunction would result in two daughter cells, one of which would be monosomic, the other trisomic. Although we cannot test for trisomy of chromosome VII, analysis of the presumptive monosomic cells for chromosome VII revealed a high coincidence of associated testable trisomy for a different chromosome. This further supported the conclusion that the white cycloheximide-resistant and leucine-requiring white colonies were indeed caused by chromosome loss due to induced non-disjunction resulting in a monosomic condition. However, this monosomic condition could have been only transient, followed by a second non-disjunctional event restoring the complete diploid condition in one of the daughter nuclei. This could explain the observation that only a few of the presumptive monosomic cells were still monosomic at the time of genetic analysis.

Chromosomal mal-segregation or non-disjunction could arise from a malfunction of the spindle fibre apparatus, which is composed in part of tubulin. Dimethylsulphoxide was shown to accelerate aggregation of bovine brain tubulin *in vitro*, and we have presented evidence here and elsewhere [5] that aprotic polar solvents do indeed interfere with tubulin aggregation *in vitro* either by accelerating aggregation (acetone) or by inhibiting it (ethyl acetate and acetonitrile). Induction of aneuploidy could be observed with both accelerating and inhibitory types of solvents. This may appear to be contradictory. However, microtubules have to be both assembled and degraded in order to move chromosomes to the spindle poles [12].

The most efficient treatment procedure resulting in observation of aneuploidy calls for an incubation period under growth conditions followed by a storage period in ice. Cultures exposed during the two periods to ethyl acetate, methyl ethyl ketone or acetone generated monosomic cells during the growth period following storage in ice [3].

It must be emphasized that induction of monosomic cells was observed without the ice-storage period. However, this ice storage amplified the effects of the treatment by a factor of ten or more. An explanation for the enhanced induction can be found in the fact that tubulin can aggregate above a certain critical temperature and dissociates at temperatures below it. That is to say that tubulin is soluble at low temperature and insoluble at high. Such a behaviour has been demonstrated also for yeast tubulin [13].

The same behaviour can be observed with macromolecules which develop a solvation sheath of water molecules. The stability of this sheath depends on the temperature: it collapses at high temperatures and regenerates at low. Once established, the solvation sheath can prevent the solvated macromolecules from aggregation. It is conceivable that aprotic polar solvents compete with water molecules for specific positions on the macromolecules. They can prevent aggregation if they bind more tightly than water to the regions involved in aggregation. Conversely, at higher concentrations they may accelerate aggregation by reducing the availability of the water. Dimethylsulphoxide, for example, was not effective at concentrations below about 9%. It is also significant that 4*M* glycerol is required in the tubulin assembly buffers.

The amplification of the induction of aneuploidy by the cold-storage step can also be explained on this basis. Microtubules formed during incubation at growth temperature in the presence of aprotic polar solvents could be weakened. They might then dissociate *in vivo* in the cold to a greater degree than without the solvents. When such cells are returned to growth temperature, microtubule formation becomes possible again, with synchronous aggregation of tubulin and associated proteins. This phase may be very sensitive to the effects of the aprotic polar solvents and a strong induction of aneuploidy will result. Cold storage is not an indispensable requirement for the induction of aneuploidy, but it may synchronize cell cultures into a phase sensitive to the induction of aneuploidy.

The proposed mechanisms of action are compatible with another peculiarity observed in the experiments reported here. Pyridine (Table 11.2) induced aneuploidy at concentrations above 0.7%, concentrations which strongly inhibited growth. The titre of colony-forming units in the control was 156.3×10^6, and a definite induction was observed at 0.79% pyridine, the titre reached being 11.9×10^6 per ml. An increase in solvent concentration to 1.28% (i.e. by a factor of 1.62) resulted in an increase by a factor of 6.4 in the frequency of aneuploidy. A similarly sharp rise in aneuploid frequency in response to a fractional increase in concentration can be seen with two lactones (Table 11.3) and phenylacetonitrile (Table 11.4). This was not observed with n-butyronitrile (Table 11.4), which induced only an increase in red resistant colonies, reflecting point mutation, mitotic recombination and/or deletion of chromosomal fragments.

Induction of mitotic aneuploidy by aprotic polar solvents shows very sharp thresholds. Even though several chemical mutagens show threshold concentrations for induction of genetic events other than aneuploidy, the thresholds usually are not so sharp as the ones observed with aprotic polar solvents [8]. Mutations were originally considered to result from one or two hits on a genetically relevant target changing the base sequence of a gene. This 'target theory' [14] predicted a linear relationship between dose and effect in the case of a single hit, and a second-order function for two hits. The very sudden response of induced aneuploidy to increasing concentrations of aprotic polar solvents further supports the conclusion that induction of aneuploidy occurs through a mechanism of action different from that of other types of genetic damage. The sudden response suggests, in analogy to the 'target theory', a requirement for a large number of 'hits'. Some critical number of target sites on tubulin molecules may be necessary for aggregation; as solvent competitively replaces water at binding sites, inhibition of aggregation finally occurs under conditions where the solvation sheath of water molecules normally collapses, the solvation sheath of e.g. methyl acetate molecules still being stable. On the other hand, increasing solvent concentrations could 'dilute' water concentrations to below the level required for the maintenance of a solvation sheath, so that aggregation is favoured and microtubules are stabilized, as seen with dimethylsulphoxide.

The concentrations of solvent necessary for aneuploidy induction were usually quite high except for phenylacetonitrile. Acetonitrile at around 4% induced aneupoloidy; however, tubulin assembly *in vitro* was inhibited at 0.1% acetonitrile concentration. This suggests that tubulins of different species could be differently susceptible to a given solvent. It may be questioned whether such solvents at the normal levels of exposure actually pose a genetic hazard to humans, but a potential hazard should be considered because several aprotic polar solvents do interfere with the formation of mammalian microtubules *in vitro* [4, 5 (Figs. 1 and 2)].

The results presented here and in another communication [3] document some points which are directly relevant to mutagenicity testing.

(1) The full mutagenic potential of a given agent can only be assessed when all possible types of genetic changes are detected in a given testing programme. Most currently used mutagenicity test batteries are restricted to effects resulting from primary lesions in DNA or resulting from the interference with enzymes involved in DNA metabolism. Other molecular targets must be considered, such as the components of the spindle fibre apparatus, tubulin, and microtubule-associated proteins. Additional structures to be considered in this context are centrioles and also the specific non-DNA and non-histone components of the kinetochore of chromosomes to which spindle fibres attach [15]. Telomeres are specialized structures at the end of chromosomes. Loss of a telomere in *Drosophila melanogaster* can lead to the loss of the chromosome [16]. If the specialized structure of yeast telomeres [17] is specifically vulnerable to a chemical, then the result might be aneuploidy. There may even be specific structures for the attachment of telomeres to the nuclear membrane that should also be considered.

(2) There are no standard dose increments when an unknown chemical is tested. Rather, the dose increments required for the detection of genetic effects have to be determined individually for each agent.

(3) There are safe concentrations, at least in the case of aprotic polar solvents, with regard to induction of aneuploidy. Many innocuous chemicals can be mutagens when applied at high enough concentrations. Any chemical which is sufficiently soluble in water to interfere with the solvation sheath of biomacro-

molecules is a potential inducer of aneuploidy. Expression of this potential depends on whether other types of damage kill the cell before the threshold concentration for interference with the spindle fibre apparatus is reached.

Acknowledgements – Part of this investigation was supported by contract no. NO1-ES-1-5005. (Development of a Yeast Aneuploidy Test System) from the National Institute of Environmental Health Sciences, Research Triangle Park, NC, U.S.A. V.W.M. was supported by a National Science Foundation Award under Agreement No. INT-8304930. Any opinions, findings and conclusions or recommendations expressed in this publication are those of the authors and do not necessarily reflect the view of the National Science Foundation. It is a great pleasure to thank Drs. H. Wolf and R. Schlegelmilch of the Institute of Pharmacology and Toxicology of the University of Ulm for suggesting that we test acetonitrile, and Dr. H.-J. Gais of the Department of Organic Chemistry of the Technische Hochschule Darmstadt for suggesting and supplying additional aprotic polar solvents. Dr. G. Schröder of Röhm GmbH, Darmstadt, pointed out to us the possible role of the solvation sheath in the temperature-dependent aggregation and dissociation of macromolecules. It was only through his help that we arrived at the postulated mechanism of action of aprotic polar solvents in the induction of aneuploidy.

REFERENCES

[1] B. J. Kilbey, M. Legator, W. Nichols and C. Ramels (eds.), *Handbook of Mutagenicity Test Procedures,* 2nd Ed., Elsevier, Amsterdam, 1984.
[2] J. M. Parry, E. M. Parry and J. C. Barrett, *Nature,* 1981, **294**, 263.
[3] F. K. Zimmermann, V. W. Mayer, I. Scheel and M. A. Resnick, *Mutation Res.,* 1985, **149**, 339.
[4] R. H. Himes, P. R. Burton and J. M. Gaito, *J. Biol. Chem.,* 1977, **252**, 6222.
[5] U. Gröschel-Stewart, V. W. Mayer, R. E. Taylor-Mayer and F. K. Zimmermann, *Mutation Res.,* 1985, **149**, 333.
[6] E. B. Hook, *Mutation Res.,* 1983, **114**, 389.
[7] K. Sankaranarayanan, *Mutation Res.,* 1979, **61**, 1.
[8] F. K. Zimmermann and I. Scheel, *Mutation Res.,* 1984, **139**, 21.
[9] R. C. Williams Jr. and J. C. Lee, in *Methods in Enzymology,* D. W. Frederiksen and L. W. Cunningham (eds.), Academic Press, New York, Vol. 102, p. 376.
[10] M. M. Bradford, *Anal. Biochem.,* 1976, **72**, 248.
[11] F. K. Zimmermann, V. W. Mayer and I. Scheel, *Mutation Res.,* 1984, **141**, 15.
[12] R. L. Margolis and L. Wilson, *Nature,* 1981, **293**, 705.
[13] J. V. Kilmartin, *Biochemistry,* 1981, **20**, 3629.
[14] C. Auerbach, *Mutation Research, Problems, Results and Perspectives,* Chapman & Hall, London, 1976.
[15] S. Brenner, D. Pepper, M. W. Berns, E. Tan and B. R. Brinkley, *J. Cell Biol.,* 1981, **91**, 95.
[16] H. J. Muller and I. H. Herskowitz, *Am. Naturalist,* 1954, **88**, 177.
[17] M. A. Forte and W. L. Fangman, *Chromosoma,* 1979, **72**, 131.

12

Ionizing radiation – a case study in the evolution of risk evaluation

P. UNRAU, Health Sciences Division, Radiation Biology Branch, Chalk River Nuclear Laboratories, Chalk River, Ontario, Canada

Ionizing radiation was first described by Roentgen in 1895 and Bequerel in 1896. Intense scientific interest in ionizing radiation arose, at least in part, because it offered a (relatively) non-invasive diagnostic method. Furthermore the researches of the Curies, and others, led directly to a deeper understanding of matter, energy, and their interrelationships. Ionizing radiation is so named because photons or particles can interact with matter, transferring energy and producing chemical ionizations. Thus a particle or wave interacts with a substrate, transferring energy to it. If the substrate is alive, the imparted energy may have effects out of all proportion to the energy input. For instance, an instantaneous total body dose of 1000 rads would be lethal to >98% of humans, yet would raise body temperature by only <0.002°C. It is the sensitivity of the target which absorbs the energy that accounts for the biological effects.

As listed in Table 12.1, an association was rapidly made between ionizing radiation and harmful effects, including skin burns (1896), radiation sickness (1897), and skin cancer (1902). Indeed, by 1922 more than 100 radiology pioneers had died of cancer or aplastic anaemia. It was this early heavy toll of prominent scientists which led directly to the establishment of the first safety standards. The later dates in Table 12.1 illustrate some of the steps that occurred during the subsequent process of standard setting. Particularly noteworthy are the founding, under the auspices of the International Congress of Radiology, of the International X-Ray Unit Committee and the International Committee on X-Ray and Radium Protection, in 1928. After reorganization in 1950 these committees persist as the International Commission on Radiation Protection (ICRP) and the International Commission on Radiation Units and

Table 12.1 – Ionizing radiation – A retrospective view

1895	Roentgen – reports X-rays.
1896	Bequerel – natural ionizing radiation from pitchblende.
	Radiation burns (erythema).
1897	Radiation sickness.
1900	Steenbeck – X-ray cure of skin tumour.
1901	Bequerel and Curie – radium-induced skin burns.
1902	Frieben – skin cancer induced.
1905	Chromosome fragmentation.
To 1921	Diagnostic uses developed.
	British X-Ray and Radium Protection Committee formed.
1922	Retrospective survey found that >100 radiology pioneers had died from cancer or aplastic anaemia.
1924	Ionization chamber developed, and doses standardized.
1925	Skin reddening (erythema) tolerance-dose concept developed (about 6 R/month).
1928	Definition of unit dose as the roentgen; 1 R = 1 electrostatic unit/cm^3.
1931	US Advisory Committee on X-Ray and Radium Protection formed.
	Tolerance dose of 0.1 R/day accepted.
1937	ICRP considered effects of doses to individual organs.
	Expanded International Commission on Radiation Protection (ICRP).
1940s	Manhattan project.
1949	National Commission for Radiation Protection (NCRP) recommended reduction of dose to individuals to 0.3 R/week.
1950	NCRP and ICRP considered risks arising from doses.
1955	ICRP recommended dose limits, considering population exposures.
	Definition of rem; unit of dose equivalent, corrected for radiation quality, tissue exposed, in man.
	Definition of rbe; comparison of *r*elative *b*iological *e*ffects of radiations of different qualities, e.g. fast neutrons and X-rays.
1960s	ICRP recommended maximum dose to public be limited to one-tenth of the limit to radiation workers.
	Linear dose response concept adopted as conservative estimate.
	Dose rate effects found in mice, suggesting lower risks from lower dose rates.
1970s	Risk concepts refined and expressed for somatic and genetic risks.
	Nuclear industry risks comparable with other 'safe' industries.
1977	ICRP proposed a dose limitation system based on (a) no practice to be adopted unless a positive net benefit accrues; (b) exposures to be kept as low as reasonably achievable, economic and social factors included; (c) doses to individuals not to exceed limits suggested by the commission.
Present	Epidemiological studies of workers in nuclear industries provoke extensive methodological debates.

Measurements (ICRU). The legacy of these committees and commissions provides a broadly based consensus on the nature of the concept of radiation protection today.

The science of genetics was coincidentally reborn and underwent extremely rapid development at the time that ionizing radiation was first discovered and utilized. Relevant developments in genetics and allied fields are summarized in Table 12.2, which is based largely on references [1] and [2]. Particularly noteworthy are the observations in 1905 that chromosomes were disrupted by ionizing radiation, and Muller's observation in 1927 of X-ray induced mutation in *Drosophila,* which was substantiated for barley by Stadler in 1928. Also, based on earlier studies by Galton, human genetics, population genetics, and quantitative genetic concepts were developed. Since that time, innumerable studies on a wide variety of organisms, including viruses, bacteria, yeast, mice and humans, have pointed out the genotoxic effects of ionizing radiation. Important fundamental biological principles such as the concept of DNA as the mutational target and the idea of its enzymatic repair, including contributions by systems inducible by radiation have been substantiated. All such developments have had an effect on radiation safety standards. For example, information obtained in such fields has been used by Muller [3] in a warning of the need to reduce population dose levels and to quantify the additional genotoxic hazards arising from nuclear war. In general however, the analysis of risks has been developed in order to generate risk/benefit data. These considerations have been greatly aided by the evolution of improved dosimetry, based at least partially on the development of faster and more sensitive assay methods.

From the definition of the roentgen (a unit describing ionization in gases) evolved the definition of the rad, a unit which defines energy deposition per gram of tissue and is therefore of much greater value for dosimetric considerations. The concept of the *rem* (*r*oentgen *e*quivalent, *m*an) was developed to ensure that units of dose were derived in terms of the radiation subject of interest, man. Additionally, target theory and microdosimetric considerations were developed and applied to the transfer of energy by radionuclides incorporated into a living cell. Indeed, the regulation of radionuclide exposures forms a subset of the regulatory approaches which may be informative to chemical genotoxicologists. For protection from radionuclides the ICRP considers the rates of ingestion, the rates of uptake, concentrations in target of source organs, doses occurring from decays within or near the target organs, and both the physical and biological half-life of the radionuclide in the target organs. The effects of several hundred radionuclides on more than thirty organs have been considered. Furthermore, the ICRP considers the decay pathways for each radioisotope and the individual contributions to the total energy given off by the various decay pathways. It is thus possible to calculate for ingested radionuclides the doses to any part of the body with a substantial degree of precision. For chemical genotoxic agents a similar approach would consider each chemical, its activation product, the ultimate degree of reactivity with DNA as a function of tissue type or cell cycle phase, the products formed, and their rates of rapair (and consequences if unrepaired). Of course, such a project is far less practical compared to the

Table 12.2 – The development of genetics – A retrospective view

1866	Mendel reported studies on inheritance in peas – ignored.
1869	Galton established some elements of population genetic studies.
1871	Miescher reported on nucleoproteins.
1870-1900	Germ plasm and developmental hypotheses.
1900	Correns, De Vries and Tschermak confirmed Mendel's observations.
1902	Genetic findings confirmed in mice and fowl. Chromosomal hypothesis proposed for genetic observations.
1903-5	Fragmentation of chromosomes by radiation.
1908	Hardy–Weinberg equilibrium equations for populations.
1921	Non-disjunction demonstrated after radiation.
1923	Crossing-over enhanced by radiation.
1927	Muller demonstrated radiation-induced mutation in *Drosophila.*
1928	Stadler demonstrated mutagenicity of X- and γ-rays in barley.
	Taken together, the data showed that neither the species nor the radiation type is specific for mutagenic damage.
	Spontaneous mutation shown to be highly temperature-sensitive.
1930	Linear dose-response curve observed for observed for radiation-induced mutation.
	No temperature effects for induced mutation.
To 1944	Ionizing radiation found to be mutagenic in most organisms.
1944	DNA implicated as genetic material.
1953	Model of DNA structure proposed by Watson and Crick.
1954	Studies of genetic effects in irradiated populations of Hiroshima and Nagasaki reported.
1955	Muller expressed concern about the effects of radiation on the genetic make-up of human populations.
1960s	Dose-rate effects noted in ionizing radiation mutagenesis studies on mice.
To 1970s	Skeletal mutations induced in mice by ionizing radiation; also with radiomimetic chemicals.
To present	Molecular basis of mutation established.
	Cloning methods developed.
	Assays for genotoxic chemicals developed.
	Molecular events leading to retinoblastoma clarified, including non-disjuunction, mutation, deletion, and somatic recombination.
	Systems approach to mutagenesis based on genetics, cytology, recombinant DNA technology, and oncogenetic changes.

relatively more easily determined numbers and nature of the products for radioisotopes.

By the 1920s the studies on X-rays and radium had spawned national bodies of experts interested in such work and concerned about the effects of radiation exposure. In turn, national bodies joined in the formation of international groups, and at these forums X-ray and radium protection could be discussed; nomenclature, dosimetry, and ultimately safety limits were standardized. The importance of both national and international committees can be judged by the fact that such bodies still exist and form the major source of advice for national authorities. Thus, at present, the ICRP makes recommendations based on expert knowledge and assessment of evidence but has no legal powers. National radiation protection groups can (and do) differ on recommendations to nationally constituted legal bodies, which set legal safety limitations and licensing arrangements.

At present, there are three major sources of information about ionizing radiation that are normally taken into account for national radiation protection standards. The United Nations Scientific Committee on the Effects of Atomic Radiation (UNSCEAR) has a predominantly scientific orientation and reports on radiation levels and effects, the latest report being in 1982 [4]. The ICRP (see ICRP 1977 [5]) primarily makes recommendations based on independent assessments, evaluations, and reports of its task forces and committees. The International Atomic Energy Agency usually publishes codes of practice and guidelines for use by developing nations. National bodies interested in radiation protection can interpret international recommendations and provide advice on particular national concerns. Politicians consider the recommendations of these bodies and have the duty to legislate regulations and safety standards. Usually the enforcement of such specialized regulations becomes the province of an independent regulatory authority within any nation.

By 1955 both national and international radiation protection bodies had become concerned with two sorts of populations; workers in the nuclear industry and the general public. The former were regulated, monitored, and protected by

Table 12.3 – Radiation exposure limitation – proposed Canadian limits

Atomic radiation workers:
- Eye lens – 150 mSv/year
- Other organs – 500 mSv/year
- Overall limit – 50 mSv/year

Members of the public:
- Eye lens – 50 mSv/year
- Other organs – 50 mSv/year
- Overall limit – 5 mSv/year ≅ 0.5 rem*

There are traditionally *no legal limits* to doses occasioned by medical diagnosis or treatment, or in the case of emergency procedures to prevent serious injury, avert danger to human life, or avert a significant loss of property.

*1 mSv ≅ 0.1 rem. Note that 5 mSv is the maximum permitted exposure to *any* individual in the population, not to all members of the public.

the dose limits then established. The general population was thereafter held to be a special case and lower dose limits were set. Table 12.3 shows the presently recommended radiation protection limits suggested by ICRP, which are being proposed for the revised Atomic Energy Control Regulations in Canada. Exposure to members of the public is generally limited to one tenth of the dose allowed to atomic radiation workers. These lower limits have arisen from concern over possible late effects in populations, with an intention on the part of the regulatory committees that radiation-induced somatic and genetic effects in the general population be limited to levels substantially lower than those of effects that can be attributed to all other natural events (i.e. significantly lower than background genetic malformation rates.)

In *dose limitations* two fundamentally different types of biological harm are considered. (1) *Stochastic effects,* such as somatic or genetic mutations, are those expected on the basis of random energy transfer in cells or in genes. They are linearly related to dose, and are expected to show no threshold. (2) *Non-stochastic effects* are also deleterious but a non-linear response curve is expected. Examples include cataracts in lens tissue, skin burns, or death. As shown in Table 12.3, the distinction between the two types of harm is clear for the case of radiation limits for the extremities, which are higher than the whole-body doses. The limits for whole-body doses include exposure of the gonads and blood-forming organs; these are more sensitive than the extremities, and irradiation of such organs may lead to long-term effects.

In practice, actual doses to *A*tomic *R*adiation *W*orkers (ARW) in nuclear research establishments are about 10% of the given limits. In nuclear power plants the doses are slightly higher. It is generally accepted by national regulatory agencies that workers in the nuclear or radiation utilizing industries may legally be subjected to higher doses than members of the general population. However, as a result of developments in cost-benefit analysis and risk determination, it has also become a general regulatory principle that radiation doses be limited so that the associated risk (e.g. of life-shortening by cancer) is as low as that in a 'safe' industry (one work-related fatality per 10^4 person-years). The regulatory recommendations also have an optimization mechanism which is known as the ALARA concept: that exposures be *A*s *L*ow *A*s *R*easonably *A*chievable, i.e. technical developments which limit doses should be accepted when socially and economically advantageous. Modifications of design, practice or technique which lead to dose reductions can thus be analysed in terms of cost versus benefit. Dose reductions resulting from design changes may most economically lead to dose reductions to workers.

Recently, a new factor has been receiving increasing attention. It has been obvious for many years that human populations are heterogeneous and may not respond to ionizing radiation as do inbred test-animal colonies or genetically uniform yeast strains. The identification of DNA-repair-defective mutants in phage, bacteria, fungi, and *Drosophila,* has led to the identification of repair-defective mutants in man. Inherited genetic repair deficiencies in humans are associated with demonstrable hypersensitivity to carcinogens and with a pre-disposition toward early cancer development (see Paterson *et al.* [6] for a recent review). In addition, since studies on lower organisms had shown that repair heterozygotes may be abnormal in their sensitivity to radiation, the subsequent statistical inference of enhanced cancer risk in human DNA-repair heterozygotes was expected. Thus while the study of human genetics to date has emphasized the subjects of population genetics and epidemiology, current research priorities include developments in the techniques of molecular biology. DNA appears to be the major sensitive target for biologically significant radiation effects and major advances in direct assays for DNA damage are likely. It is expected that assays based on HPLC and mass spectroscopy (see, for example, McLafferty [7]) or using damage-specific monoclonal antibodies and specialized detection techniques such as ELISA (cf. [8]) will make possible the detection of the extremely small (femtomole) quantities of damaged bases in cells exposed to carcinogens, i.e. a few damaged bases per cell will be detected. Other methods such as the search by Neel *et al.* [9] for mutations amongst 28 blood and serum enzymes, by use of electrophoretic techniques (many additional proteins and thus genetic loci are amenable to image-enhancement mapping techniques), are likely to lead to a more precise determination of the relationship between mutagenic effects and genotoxic risks of ionizing radiation or suspect environmental chemicals. It is therefore probable that within a reasonable time span direct measurements of genotoxic harm will be established which will enable direct comparisons between various agents.

Table 12.4 gives highlights in the history of chemical genotoxicology. The earlier history of ionizing radiation is briefly recapitulated, with the formation of national and international expert committees. Note that the development of biological assays of genotoxicity was crucial: in the case of chemical agents, microbial genetics led the way to the development of bacterial, yeast and other assays which together appear to detect >95% of all known carcinogens. Further, developments in epidemiology and studies of selected populations which eschew carcinogenic habits such as smoking or coffee-drinking may lead to the identification of chemical carcinogens in the workplace or the home.

Bridges [10] suggested one possible approach concerning the development of safety standards for chemical genotoxic agents. He proposed the establishment of assessment systems designed to give the most effective screening of critical chemicals, with a view to their elimination, substitution, or strict control. Furthermore, he suggested that chemical systems be cross-standardized to ionizing radiation, to standardize risks. This 'rad equivalent' concept has not yet been followed up, but improved techniques of biochemical analysis may allow an extension of this idea at some future date. A practical equivalence could be based on the yields of modified DNA bases (e.g. as determined by mass spectrometry), thus establishing a *modified base equivalent dose* for each agent. Genotoxic agents that result in modified bases could then be intercompared for both the amounts and types of base damage per unit dose.

The polyimides may be used to illustrate the application of some of these principles. The many attractive physical and chemical properties of polyimides are countered by the significant mutagenicity, and therefore risk to workers, of the diamines prior to polymerization. However, it has proved possible to modify

Table 12.4 – The development of genotoxicology – A retrospective view

1775	Potts characterized scrotal cancer as an occupational disease of chimney sweeps.
1915	Mustard gas and other chemicals used in warfare. Similarities noted between radiation and mustard gas effects, including slow healing, skin burns, etc.
1940–41	Mustard gas shown to be mutagenic in *Drosophila.*
1947	Studies on chemical mutagens published.
To 1960s	One gene–one enzyme hypothesis established.
	The nature of mutational changes induced by chemicals established. Relationships between smoking and lung cancer established.
1960s	Mutagens and recombinagens related to carcinogenesis.
	Development of test systems in phage, bacteria, yeast and higher eukaryotes.
1969	Mutagenic ('error-prone') repair reported; cross-sensitivity of chemical and radiation sensitive mutants reported.
	Environmental Mutagenesis Society founded.
1970s	Ames test for mutagenic activity of chemical agents achieves broad acceptance.
1974	Radiation equivalent dose suggested as a basis for later regulation of carcinogenic chemicals.
1977	Foundation of International Commission for Protection against Environmental Mutagens and Carcinogens.
1979	Environmental Mutagenesis Journal begins publication.
Future	Definition of carcinogenic units.
	Systems for dose limitations of known carcinogens should be established.
	Development of methods to measure chemical damage per base pair.

the diamines so that they are not detectably mutagenic, though still retaining the features that make them attractive subunits for polymerization. Furthermore, a combined mutagenesis/development programme has led to improved understanding of the mutagenic nature of diamines; this approach has been described by St. Clair and Johnston [11].

It can be confidently predicted that disagreements of the sort that are at present generated by radiation protection standards will also greet future safety standards for chemical genotoxic agents. Interminable recitations of the current carcinogen of the week must not dull our realization that significantly carcinogenic chemicals do exist. It is also expected that individual humans will be identified who are sensitive to chemicals for genetic reasons, and that current concern and discussion about heterogeneity and repair heterozygotes in the radiation context will enlighten attitudes toward future safety standards of genotoxic chemicals.

The full history of ionizing radiation safety standards has not yet been written. To date it has proved to be an activity in which a network of information has been established, which has been used to set acceptable limits. This network is constantly evolving and is necessarily incomplete, because no area of radiation biology and jurisprudence has yet been exhausted.

It is expected that the regulation of safety standards for chemical genotoxic agents will similarly arise from a network of chemical, biological, and epidemiological knowledge. It is hoped that the resulting network will detect and help control carcinogenic effects of chemicals without unduly restricting technical developments. Control of ionizing radiation hazards provides a model of a balance between accepting the benefits of technology and controlling the attendant risks.

Acknowledgements – Some tables include information from an unpublished historic outline supplied by A. M. Marko. D. K. Myers and R. E. J. Mitchel provided suggestions for additions to the text. The editorial assistance of N. E. Gentner is gratefully acknowledged.

REFERENCES

[1] A. H. Sturtevant, *A History of Genetics,* Harper and Row, New York, 1965, pp. 69–72, 136–138.
[2] H. L. K. Whitehouse, *Toward an Understanding of the Mechanism of Heredity,* Macmillan, Toronto, 1969, pp. 136–144.
[3] H. J. Muller, *J. Heredity,* 1955, **46,** 199; *Sci. Am.,* 1955, **193,** 58.
[4] ICRP No. 26, *Recommendations of the International Commission on Radiological Protection,* Pergamon Press, Oxford, 1977.
[5] UNSCEAR (1982), *Ionizing Radiation: Sources and Biological Effects,* United Nations Scientific Committee on the Effects of Atomic Radiation, U.N., New York, 1982.
[6] M. C. Paterson, N. E. Gentner, M. V. Middlestadt and M. Weinfeld, *J. Cellular Physiol. Supplement,* 1984, **3,** 45.
[7] F. W. McLafferty, *Science,* 1981, **214,** 280.
[8] G. R. Mohn, P. R. M. Kerkloan, A. A. van Zeeland, J. Ellenberger, R. A. Boan, P. H. M. Lohman and F. W. Pons, *Mutation Res.,* 1984, **125,** 153.
[9] J. V. Neel, C. Satoh, H. B. Hamilton, M. Otake, K. Goriki, T. Kageoka, M. Fujita, S. Neriishi and J. Asakawa, *Proc. Natl. Acad. Sci. U.S.A.,* 1980, **77,** 4221.
[10] B. A. Bridges, *Mutation Res.,* 1974, **26,** 325.
[11] T. L. St. Clair and N. T. Johnston, *Symposium on Advanced Materials Technology,* NASA, Washington, DC, N83-12147, 1982, pp. 29–48.

Index